The Reader's Digest

# GOOD BEACH GUIDE 1997

**with Ordnance Survey Maps**

COMPILED BY THE
**MARINE CONSERVATION SOCIETY**

with a foreword by
**Jonathon Porritt**

David & Charles

*Acknowledgements*
This guide was compiled with information supplied by the Environment Agency in England and Wales, the Scottish Environmental Protection Agency, the Department of the Environment in Northern Ireland, the Governments of Jersey, Guernsey and the Isle of Man, the private water service companies, the Water and Sewerage sections of the Regional Councils of Scotland, Environmental Health Offices, Tourism Offices, the National Trust, and data from the Reader's Digest Beachwatch survey which was co-ordinated by the Marine Conservation Society. All information received up until 1 January 1997 has been included. Neither the Publishers nor the Marine Conservation Society can take any responsibility for changes occurring after this date.

The Editors would like to thank all those who provided information, Reader's Digest for sponsoring the publication and the staff and volunteers of the Marine Conservation Society for support and assistance.
*Mark Gibson, Guy Linley-Adams, Sam Pollard, Marine Conservation Society*

*Picture Credits*
Paul Watts p7, 21, 24, 29, 52; Chris Warren p9, 166, 169, 175; Paul Glendell 14, 34; Parkin Estates Limited p16; Hartland Quay Hotel p19; T Dingle p22; Carrick District Council p26; Restormel Borough Council p33; South Hams District Council p37, 39; Torbay Tourist Board p41; English Riviera Tourist Board p42, 45, 46; Glider Advertising p49; South West Water p51; Graham Sapsford p55; Bill Foster/MCS p57; R Jarman 58–9; Chris Newton/MCS p60; J R Cox p63; Tourism Services, Poole p64; Bournemouth Tourism & Publicity Bureau p67; Sarah Wellon/ MCS p69; Havant Borough Council p84, 91; NFDC – Leisure, Tourism p87; Thanet District Council p97; Swale Borough Council p98, 101; North Norfolk District Council p104; Borough Council of Kings Lynn and West Norfolk p107; Berwick-upon-Tweed Borough Council p114; Alnwick District Council p120, 123; Sue Gubby/MCS p129; Scottish Nuclear Ltd 134–5; East Lothian District Council p136–7; STB/Still Moving Picture Co p138–9; Balmedie Country Park p142–3; Eric Ellington, SNH p145, 146–7; Isle of Anglesey County Council p157; Wales Tourist Board p159, 163; Preseli Pembrokeshire District Council p171, 172; South Pembrokeshire District Council p181; Pembrokeshire County Council p182; Camarthenshire County Council p184; Swansea Country Council p187; Ogwr Borough Council p188, 189; Limavady Borough Council p203; Mike Williams p204; Simon Boyle, Delamont Country Park p205; Northern Ireland Tourist Board p207, 209; Stuart Abraham p212; Guernsey States Board of Administration p215, 217, 218–9, 221, 223, 224–5; Peter McMahon p227; Jersey Tourism p229, 233, 235, 237, 240; Tom Mackie p234, 239; Ethel Davis p238.

A David & Charles Book
Copyright © Marine Conservation Society, David & Charles, 1997
Maps reproduced from Ordnance Survey Landranger 1:50,000 series with the permission of Her Majesty's Stationery Office © Crown Copyright, Permit No 768 – Northern Ireland

First published 1997
Tenth annual edition

All rights reserved. No part of this publication may be reproduced, stored in a retrieval system, or transmitted, in any form or by any means, electronic or mechanical, by photocopying, recording or otherwise, without prior permission in writing from the publisher.

A catalogue record for this book is available from the British Library.

ISBN 0 7153 0488 7 Paperback
ISBN 0 7153 0496 8 Hardback

Printed by Butler & Tanner Ltd, Frome
for David & Charles
Brunel House, Newton Abbot, Devon

# FOREWORD

Enjoyment of the coast is still very much a part of the British summer, though these days people are as likely to don a wet-suit and grab a surfboard or sub-aqua kit as they are to recline in a deck-chair or stroll along the promenade. Nor do we stop visiting the coast as soon as the summer is over; surfers, windsurfers, divers and others are on the beach and in the water all year round.

While the leisure activities we choose to pursue at the beach may have changed, some features of the British seaside remain depressingly familiar. Untreated sewage, for instance, continues to be discharged into the sea much as it was at the turn of the century: millions of litres of raw or inadequately treated sewage flow into our coastal waters every day.

Since the beginning of the 1990s vast sums have been invested in schemes to enhance water quality at British bathing beaches through the improved treatment of sewage. This move to clean up our seas results, in part, from the UK's obligations under European legislation. Equally significant, however, has been the role played by consumer pressure. In response to this pressure, companies such as Welsh Water and now Wessex Water have realised the benefits of a progressive attitude to sewage pollution control. But progress is very slow. The figures show that there has been no real improvement over the past year and there is some suggestion that water companies are not investing the money they should.

Only by having comprehensive and accurate information on the state of Britain's beaches can we, as consumers, bring pressure to bear. Never underestimate the impact of more and more of us 'voting with our feet', avoiding those beaches that do not make the grade – where bathing may make us ill – and frequenting only those where water quality comes up to scratch. For a decade now, a unique publication has been the most authoritative source of this information.

The Reader's Digest *Good Beach Guide*, compiled by the Marine Conservation Society, tells us everything we need to know about water quality and sewage discharges at beaches around England, Northern Ireland, Scotland, Wales, the Channel Isles and the Isle of Man. It sings the praises of the best beaches, and it is not afraid to identify those stretches of our coast which are still grossly affected by sewage and which need cleaning up. The *Guide* also provides the kind of advice that empowers us all to address the problem of dirty beaches. Information such as where to complain when you come across a polluted beach, or how we can all help reduce our personal contribution to the problem of sewage-derived litter, by supporting campaigns such as 'Bag It and Bin It' and not flushing sanitary items down the toilet.

We all benefit when our beaches and coastal waters are clean and free from pollution. Dirty beaches and seas dissuade tourists from visiting and are a drain on local economies. They present a threat to wildlife and to human health. This is not the time to think the battle is won: we must go on demanding improvements in the quality of bathing waters, improvements that result in more than just bare minimum standards being met. If this sounds like good sense, then help the Marine Conservation Society continue the fight: use the *Good Beach Guide*, avoid the polluted beaches, and shame the government and the water companies into action.

*Jonathan Porritt*

# 1:50 000 Landranger Series Map
# CONVENTIONAL SIGNS

## Ordnance Survey

### ROADS AND PATHS   Not necessarily rights of way

| Symbol | Description |
|---|---|
| Service area M1, Junction number 3, Elevated | Motorway (dual carriageway) |
| | Motorway under construction |
| Unfenced, Footbridge, A 40 (T), Dual carriageway | Trunk road |
| | Main road |
| | Main road under construction |
| B 284 | Secondary road |
| A 855, Bridge, B 885 | Narrow road with passing places |
| | Road generally more than 4m wide |
| | Road generally less than 4m wide |
| | Other road, drive or track |
| | Path |
| | Gradient: 1 in 5 and steeper, 1 in 7 to 1 in 5 |
| | Gates    Road Tunnel |
| Ferry P, Ferry V | Ferry (passenger)   Ferry (vehicle) |

### RAILWAYS

- Track multiple or single
- Track narrow gauge
- Freight line, siding or tramway
- Station  (a) principal  (b) closed to passengers
- Level crossing LC
- Embankment
- Cutting
- Bridges, Footbridge
- Tunnel
- Viaduct

### ROCK FEATURES

outcrop, cliff, scree

### PUBLIC RIGHTS OF WAY   (Not applicable to Scotland)

- ............... Footpath
- – – – – – Bridleway
- –·–·–·– Road used as public path
- +·+·+·+ Byway open to all traffic

Public rights of way indicated by these symbols have been derived from Definitive Maps as amended by later enactments or instruments held by Ordnance Survey on     (date)     and are shown subject to the limitations imposed by the scale of mapping. Later information may be obtained from the appropriate County or London Borough Council

The representation on this map of any other road, track or path is no evidence of a right of way

**Danger Area**   Firing and Test Ranges in the area.
Danger! Observe warning notices

### HEIGHTS

Contours are at 10 metres vertical interval
144   Heights are to the nearest metre above mean sea level

Heights shown close to a triangulation pillar refer to the station height at ground level and not necessarily to the summit.

1 metre = 3.2808 feet

### GENERAL FEATURES

- ruin   Buildings
- Public buildings (selected)
- Quarry
- Spoil heap, refuse tip or dump
- Coniferous wood
- Non-coniferous wood
- Mixed wood
- Orchard
- Park or ornamental grounds

- Electricity transmission line (with pylons spaced conventionally)
- Pipe line (arrow indicates direction of flow)
- Radio or TV mast
- Places of Worship: with tower / with spire, minaret or dome / without such additions
- Chimney or tower
- Glasshouse
- Graticule intersection at 5' intervals
- Heliport
- Triangulation pillar
- Windmill with or without sails
- Windpump/Wind Generator

### WATER FEATURES

- Marsh or salting
- Lake
- Canal, lock and towpath
- Canal (dry)
- Aqueduct
- Footbridge
- Normal tidal limit
- Lighthouse (in use and disused)
- Beacon
- Slopes
- Cliff
- Flat rock
- Low water mark
- High water mark
- Mud
- Sand
- Dunes
- Shingle

### ABBREVIATIONS

- P    Post office
- PH   Public house
- MS   Milestone
- MP   Milepost
- CH   Clubhouse
- PC   Public convenience (in rural areas)
- TH   Town Hall, Guildhall or equivalent
- CG   Coastguard

### ANTIQUITIES

- VILLA   Roman
- Castle   Non-Roman
- ⚔ 1066   Battlefield (with date)
- ∴   Tumulus
- +   Position of antiquity which cannot be drawn to scale

The revision date of archaeological information varies over the sheet

### BOUNDARIES

- –+–+–+– National
- –○–○–○– London Borough
- –+–+–+– District
- NT   National Trust
- NT   always open
- NT   limited access, observe local signs

NTS (in red or blue) National Trust for Scotland

- County, Region or Islands Area
- National Park or Forest Park

### TOURIST INFORMATION

- ℹ    Information centre, all year/seasonal
- Selected places of tourist interest
- Viewpoint
- P    Parking
- ⚔    Picnic site
- ⛺    Camp site
- Caravan site
- ▲    Youth hostel
- Golf course or links
- Bus or coach station
- 📞    Public telephone
- Motoring organisation telephone
- PC   Public convenience (in rural areas)

## Scale 1:50 000

Kilometres   0  1  2  3  4
Statute miles   0  1  2

# Contents

Acknowledgements **2**

Foreword **3**

Key to map symbols **4**

INTRODUCTION
**6**

HOW TO USE THE GUIDE
**10**

SOUTH-WEST ENGLAND
**15**

SOUTH-EAST ENGLAND
**85**

THE EAST COAST
**115**

NORTH-WEST ENGLAND AND THE ISLE OF MAN
**129**

SCOTLAND
**133**

WALES
**155**

NORTHERN IRELAND
**201**

THE CHANNEL ISLANDS
**213**

BRITAIN'S COASTLINE UNDER THREAT
**244**

Appendix and Index **250**

# INTRODUCTION

As a nation of island dwellers, the British possess a long-held affinity for the sea. The UK has around 15,000 kilometres of coastline – the longest in the European Union – ranging from dramatic cliffs to seemingly endless sand dunes. With relatively easy access to coastal areas (no point in the country is more than 200 kilometres from the sea) most of us have visited the coast at one time or another. All too often, however, we imagine the seas around Britain as lifeless and forbidding, when the reality is very different. Our coastal waters contain as rich and diverse an array of marine life as one could wish to find; although cold, they are very productive, supporting populations of everything from exquisite anemones to majestic basking sharks, from luxuriant kelp forests to playful dolphins. In addition to 'home-grown' wildlife, such as seabirds and grey seals, the British coast is of international importance for over-wintering and migrating birds. It is the responsibility of us all to treat our coast and coastal waters with the respect they deserve, and to ensure we keep the marine environment fit for wildlife and for our own use.

And yet the seas around our coasts are under constant pressure. Each day hundreds of millions of litres of raw or partially treated sewage are pumped into the sea, affecting marine life, causing us to fall ill and reducing some of our finest stretches of coast to an aesthetically revolting mess. Much of this sewage – which includes not just human waste but everything else that gets flushed down the toilet: sanitary protection, condoms and all manner of bathroom junk – ends up on our beaches. As a result the shores around Britain are littered with sewage-derived rubbish. A recent survey conducted by the National Rivers Authority (now part of the Environment Agency) in the south-west of England showed that out of 202 beaches only eight were judged to be free of sewage-related debris and many were found to be objectionable. The 1996 Reader's Digest Beachwatch survey, organised by Marine Conservation Society, reported that up to 45 items of sewage related debris per metre were found on some beaches in the north-west of England. Littering of this sort on such a huge scale is visually repulsive, but it is more than just an aesthetic problem. Plastic is often harmful to wildlife: many seabirds are known to ingest it, mistaking it for food. The scale of the problem is graphically demonstrated by the fact that 80 per cent of the gannet nests on Grassholm, the third largest colony in the world, contain plastic in one form or another.

The quality of the beaches and bathing waters around our coasts has been the subject of controversy and heated debate for the past four decades. Initially, the lack of an independent and impartial source of information on water quality presented a major problem. This was addressed by the publication of the first Golden List of Beaches, compiled by Tony and Daphne Wakefield following the death of their daughter, Caroline, from polio which they believe she contracted from swimming in sewage-contaminated water. The Wakefields established the Coastal Anti-Pollution League (CAPL) to campaign against the disposal of raw sewage near bathing waters. In the 1980s CAPL merged with the Marine Conservation Society, an environmental organisation working exclusively to safeguard the marine environment. Over the years the Marine Conservation Society has continued and expanded the work of CAPL and has campaigned for improvements in the treatment of our coast and coastal

*Take only photos, leave only footprints – advice applicable to Constantine Bay or any part of our coast.*

waters, particularly with regard to sewage pollution. This guide is the tenth produced by the Marine Conservation Society and is directly descended from the first Golden List of Beaches.

There is much conflicting information about the state of Britain's beaches. Their eligibility for a range of different awards and accolades, many of which mean different things, only adds to the confusion. Two of these, the European Blue Flag and the Seaside Award (a yellow and blue flag), are regularly given to beaches with widely differing water quality. Beaches flying the European Blue Flag have met the Guideline standards (*see page 11*) of the EC Bathing Water Directive in the previous bathing season and should therefore have high quality bathing water. But the Seaside Award, introduced for largely political reasons when the Blue Flag standards were tightened up, and managed by the Tidy Britain Group, can be awarded to beaches with far lower water quality, in some cases the bare legal minimum allowed by the Directive. A beach flying the Seaside Award may be well managed and have good facilities; the water quality, however, is not guaranteed. We at the Marine Conservation Society would like to see suitably stringent water-quality criteria introduced into the award of flags of any colour. The cleanliness of the water is among the most important considerations when visiting a beach, and the giving of awards to beaches with sub-standard water quality devalues the whole system.

Precisely because so much of what we read about the state of the water at our beaches is conflicting and ambiguous it is essential to have an independent and impartial source of information, and that is what this book aims to provide. The only definitive guide to the quality of bathing water in Britain, it has been compiled to give a clear guide to the state of our beaches and to allow you, the reader, to make an informed choice about where – and whether – to bathe. Britain has thousands of beaches and over 1000 of them are included and summarised here. The very best are featured in detail with photographs and maps of the area.

We all have our ideas about what makes a 'good' beach; the prime criterion for a beach to be featured here, however, is water quality. If you are wondering why your favourite beach is not included in the *Guide*, the answer may be that the water quality is not monitored, or that monitoring has shown that the quality of the sea water is not satisfactory, or possibly that inadequately treated sewage is discharged there. Remember, even a beach that appears superficially immaculate may have dangerously high levels of bacteria and viruses in the water due to sewage contamination.

Although it is the prime reason that this book was written, sewage is not the only problem to plague the coastline. It is also under pressure from construction, industry and fishing. Fortunately, some areas do enjoy a degree of protection. Britain has around 40 stretches of Heritage Coast, unspoilt areas with high scenic value. Although this is a definition rather than a statutory designation, specific management plans and policies exist for these sites to ensure their survival. There are many National Scenic Areas and Areas of Outstanding Natural Beauty in coastal locations; these too are protected through management plans and through local planning policies. In addition, there are the statutory designations – numerous coastal areas have been identified as Sites of Special Scientific Interest because of their flora, fauna and landscape features. These are managed through agreements with the nature conservation agencies of England, Scotland and Wales, which restrict potentially damaging operations.

Many terrestrial nature reserves have coastal sections and through management to protect the terrestrial environment, the coast too is protected. The National Trust, through a project called Enterprise Neptune, owns nearly 900 kilometres of coastline in England, Wales and Northern Ireland. Many of the beaches featured in the *Guide* are in the care, or under direct management, of the National Trust; at others, the land adjoining the beach is owned by the National Trust and access for visitors has been improved. The National Trust also manages conservation interests such as sand dune restoration, or cliff grazing schemes to maintain habitats for coastal birds and other species.

The sites and reserves described above are for the protection and management of coastal land. There is, by contrast, very little protection for the environment below the surface of the water. The UK has only three statutory Marine Nature Reserves, which protect wildlife inhabiting the waters and sea-bed around Lundy, Skomer and within Strangford Lough. A fourth is proposed around the Menai Straits, North Wales. In all these cases the total area of protected marine environment is relatively small, no more than a few hundred square kilometres. These reserves were designated because they are known areas of interest, but much more of the marine environment deserves the same level of protection. And even then, legal protection does not always mean real protection: Skomer was badly polluted by oil from the Sea Empress oil tanker disaster in 1996.

Thankfully, the imminent introduction of the EC Habitats Directive will mean improved protection for UK marine sites. This piece of Europe-wide nature conservation legislation is designed to ensure that areas containing important examples of species or habitats considered to be particularly rare, endangered or vulnerable are not adversely threatened by human activities. In Britain, this will result in the establishment of a number of marine Special Areas of Conservation (SAC), each of which will be managed with the protection of specific types of habitats and species in mind. Ultimately there may be as many as 40 SACs around the British Isles, covering up to 20% of our coastal waters.

The oceans may be huge, but it is the coastal areas and the continental shelves that are the most productive and most at risk. Conservation initiatives such as those described above are welcome but still we continue to exploit the sea in ways and to extents which may no longer be sustainable. This is why the marine environment needs comprehensive management and protection as much as any site on land. Just because we cannot see the effects doesn't mean that all is well below the sea surface: scenes of unspoilt tranquility may mask a situation where fish stocks are close to collapse, or life on the sea bed is decimated by pollution.

Few of us get the chance to explore the sea bed, but wherever we go, on beach or rocky shore, we should be thoughtful and responsible when exploring the coast. Visitor pressure can disrupt shore life and lead to the erosion of fragile ecosystems. When at the seaside, follow the Marine Conservation Society's Seashore Code: be careful at the coast, always keep an eye on the tide and be aware that the sea can be a very hostile environment. Show respect for the coast: don't drop litter, drive on the roads, not on the beaches, and try to avoid disturbing the wildlife.

Take nothing but photos, leave nothing but footprints, waste nothing but time.

*The Norman castle at Manorbier (p183) has splendid views to the bay and beyond.*

# How to use the Guide

This chapter includes a key to the water quality rating system used in the Guide and an explanation of the summary information presented in the tables. It will enable you to plan your holiday around one of the featured beaches, or possibly one of the listed beaches with top-rated water quality and an adequate level of sewage treatment (a full explanation of sewage treatment and its implications for your health is given on pages 244 - 249). If you have already chosen your destination, the Guide will tell you whether or not it is considered safe to swim when you get there.

**Featured Beaches and Summary Listings**
The Guide is divided into eight regional chapters: the South-West, the South-East, the East Coast, the North-West and the Isle of Man, Scotland, Wales, Northern Ireland and the Channel Islands. Each chapter lists all the beaches identified under the EC Bathing Water Directive (see below) and many non-identified beaches – over 1000 in total. All but one of these regions have beaches that we are happy to recommend, with a high standard of water quality and a low probability of contamination from sewage, and these beaches are fully featured with photographs, maps and a dossier of useful information. The summary listings of beaches that do not, for one reason or another, merit recommended status – because of unsatisfactory water quality or level of sewage treatment, a litter problem, or perhaps due to lack of adequate information – give a water quality rating for each beach, sewage outfall details and explanatory notes, all of which will enable you to decide for yourself whether or not you want to visit these beaches.

**The EC Bathing Water Directive (76/160/EEC)**
We make no apology for regarding water quality as the most important feature of any beach. The basic quality standard is laid down by the European Commission in the EC Bathing Water Directive (76/160/EEC). The Directive does not have a scientific base, however, and it is widely felt that the microbiological standards set are rather arbitrary. We at the Marine Conservation Society take a much more rigorous scientific approach, and for that reason we insist that beaches must reach a significantly higher standard before they are featured in the *Good Beach Guide*.

At present around 472 bathing waters have been identified in the UK under the EC Bathing Water Directive. These include all featured beaches, except where noted, and other listed beaches indicated by the letters 'EU'. These sites are regularly monitored by the Environment Agency in England and Wales, the Scottish Environmental Protection Agency and the Department of the Environment in Northern Ireland (DoE - NI). Samples of seawater are taken from the identified areas at regular intervals during the official bathing season; these samples are analysed for the presence of coliform bacteria, faecal coliform bacteria and faecal streptococcus bacteria, and the counts are recorded. All of these bacteria are found in the human gut and are therefore useful indicator species for sewage pollution. Some are better indicators than others: coliform bacteria can occur naturally in the environment and may not necessarily have human origins. Faecal streptococcus bacteria, on the other hand, are almost always associated with human sewage, and their presence in a sample is unambiguous. These bacteria can cause illness, especially by infection through cuts and wounds. Other pathogens regularly present in sewage include enteric viruses, salmonella and the hepatitis A virus. Although the Directive allows for the monitoring of enteric viruses and salmonella, the Government bases its results only on the coliform and streptococcus counts and we are therefore limited here to using data relating to these specific indicators.

**The Mandatory Standards and the Guideline Standards**
The EC Bathing Water Directive sets out two standards against which water cleanliness is measured: the Mandatory Standards (also known as the Imperative or Minimum Standards) and the Guideline Standards which are twenty times stricter. The Directive stipulates that member states *must comply* with the Mandatory Standards and should *strive to achieve* the Guideline Standards. Nearly all Britain's identified bathing beaches were to have reached the Mandatory Standards – i.e. the minimum legal standards – by the end of 1985. In 1991, in consultation with the European Court, the government pledged to ensure that this modest ambition would be achieved by the 1995 bathing season. But once again our beaches failed to make the grade, with 11% of bathing waters failing the minimum standards in 1996. When a beach is declared a pass or fail by the Department of the Environment, it is the Mandatory Standards that are referred to – it should be noted that a pass at this standard is not a claim to be sewage free, merely that the sewage has reached a certain degree of dilution.

The European Commission has put forward proposals to amend the Directive. These amendments would represent a considerable tightening of the present standards, although the Marine Conservation Society still feels they do not go far enough. The amended Directive was due to come into force at the end of 1995, but opposition to the tightening of the standards from, among others, the UK Government and the Water Services Association means that tighter standards are unlikely to be agreed by the EU this side of the year 2000. Meanwhile the Marine Conservation Society will continue its campaign to bring about a much more rapid strengthening of the law.

An increasing amount of scientific evidence shows that those beaches passing only the Mandatory Standards present a health hazard to bathers. Under certain circumstances, bathing waters containing low levels of bacteria used as indicators of sewage pollution may still harbour significant quantities of other sewage-derived organisms, and these carry health risks of their own. The only sure way to reduce such a risk is to ensure that inadequately treated sewage is not allowed to contaminate our recreational waters. By treating sewage to a secondary level (see page 244) up to 98% of the micro-organisms originally present in the effluent are removed. This dramatically reduces the risk of bathing water being routinely contaminated with sewage-related bacteria and viruses that may ultimately be harmful to health.

**The New Marine Conservation Society Water Quality Grades for 1997**
For the tenth edition of the Guide, the Marine Conservation Society has divided bathing waters into three categories (not five as in the past), simplifying the system and making it easier to understand whether a beach is recommended or not. The new grades are:

*How to use this book*

### F – Fail
*less than 95% pass of the Mandatory Standards*
The Department of the Environment regards this as a fail. These beaches are heavily contaminated by sewage. The Marine Conservation Society advises that these waters should not be used for bathing or any other water contact sports.

### P – Mandatory Pass
*95% pass of the Mandatory Standards*
The Department of the Environment regards this as a pass of the mandatory standards, the minimum legal requirement for bathing beaches. Waters just passing the mandatory standards are almost certainly contaminated by sewage and carry a significant health risk according to recent research carried out by the Water Research Centre. Some bathing waters in this group may have achieved the bare minimum legal requirements quite easily, but may not be clean enough to be considered for recommendation. It should be especially noted that some beaches may be awarded a Guideline pass by the Government while only achieving 95% compliance with the Mandatory Standards coupled with the specific requirements for a Guideline pass. The Marine Conservation Society believes that 95% compliance with the Mandatory Standards is inadequate and will not recommend a beach if it has failed the Mandatory Standards at any time over the bathing season.

###  – The Marine Conservation Society Guideline Pass
*100% pass of Mandatory Standards,*
*80% pass of Guideline Coliform Standards, or better*
*90% pass of Guideline Faecal Streptococcus Standards, or better*
These beaches have the lowest bacterial counts and cleanest water in the country, and as such can be considered for recommendation in the *Good Beach Guide*. The Marine Conservation Society is satisfied that bathing in these waters represents a minimal risk to health.

The Marine Conservation Society will therefore only recommend those beaches which:
1) have obtained a Marine Conservation Society Guideline Pass, and;
2) are not affected by any sewage outfalls, unless the discharge is treated to at least secondary level.

To summarise then, the three categories of water quality are:
**F** – illegally polluted, avoid this beach
**P** – passes minimum legal requirements, but still not clean enough
**G** – clean enough to be considered for recommendation by the Reader's Digest *Good Beach Guide*.
The symbol '-' indicates that no sampling was undertaken at this beach, or no data was available.

### Track Record
Most beaches include a record of which water quality category (F, P or G) they achieved over the preceding four years; for example, '-FPG' would indicate that the beach obtained a Marine Conservation Society guideline pass in 1996, a mandatory pass in 1995, that it failed

in 1994 and that no sampling data was available in 1993. When a beach, even a recommended beach, has a record of sporadic water quality, it is a good idea to check with the Environment Agency or local environmental health department for the very latest water quality data. The track record covers the years 1993-1996 inclusive.

**Sewage Discharges and Sewage Treatment**
The Sewage Outlet column of the tables contains detailed information on the number of sewage outlets in the area, the degree of treatment the sewage has undergone prior to discharge, the number of people served by the outlet and where the discharge point is relative to low water mean** (all distances are given in metres unless otherwise stated). For example, the entry: *1, screened and disinfected, 75,000, 100 below LWM* means one outlet serving 75,000 people discharges screened and disinfected sewage 100 metres below the low water mark. This level of detail is available nowhere else and, taken in conjunction with the water quality grade and the track record, gives you the fullest possible information about a given beach, enabling you to make an informed choice about where or whether to bathe. Information is provided by organisations including the Environment Agency, SEPA, DoE-NI and the Water Companies and was correct at the time of going to press.

** For Wales, most discharge points are given relative to the horizontal distance below the high water spring tide mark (HWST).

**Safety at the Seaside**
Please take care at the beach. We have tried as far as possible to note those areas where it is dangerous to bathe, but remember that the sea is a highly variable environment which under certain conditions can be extremely hostile. There are some basic rules to follow when bathing at the seaside to minimise the possibility of getting into trouble in water. Always listen to lifeguards, follow their advice and make sure that you understand the system of safety flags; do not swim when the sea is rough or where there are known currents or riptides; swim parallel to the shore rather than out to sea; don't swim immediately after a meal and never after drinking alcohol. If you see someone else in trouble, fetch the lifeguards or contact the Coastguard (999) – do not attempt to rescue them yourself. Further information of lifesaving and training schemes is available from the Royal Life Saving Society whose address is given in the Appendix.

**Maps**
All mapping in the *Guide* is from the Ordnance Survey Landranger 1:50,000 series and is reproduced at this scale. Each two-centimetre square represents one kilometre.

**Abbreviations**
The following abbreviations are used throughout:
LWM – low water mark
HWM – high water mark
UV – ultraviolet
LSO – long sea outfall (see page 245)
HWST – high water spring tide mark

*The South-West is home to many of Britain's cleanest beaches, such as this one at Challaborough (p34).*

# South-West England

Covering the coast from Clevedon near Bristol to Bournemouth in Dorset, and including the Isle of Wight, this section encompasses some of the best beaches in the country and some of the foremost resorts. The coastal scenery ranges from the heavily industrialised areas around the Severn Estuary to the rugged beauty of the north Cornish coast.

•

There are hundreds of kilometres of glorious coastline along the south-west peninsula and the south-west coast has a high number of bathing waters identified under the EC Bathing Water Directive. Long sweeping bays and small secluded coves are separated by rugged headlands. Spectacular rocky cliffs contrast with smooth turf slopes where wild flowers abound. Some of Britain's loveliest unspoilt scenery is to be found along this coast and many of the cleanest beaches and bathing waters are located in the West Country.

The area is not free of problems, however, with various forms of pollution affecting a number of beaches. Untreated sewage is discharged close inshore and is washed back on to the sands at popular resorts. Both South West Water and Wessex Water have considerable investment programmes to deal with the sewage problems of the region but there is still much to do; in the short term, it appears that sewage-related debris will remain an everyday obstacle faced by swimmers, surfers and divers at many popular beaches. On a more positive note, Wessex Water has recently committed itself publicly to ensuring secondary treatment of all sewage discharges in its sector of the south-west.

Severe congestion builds up in the summer as large numbers of tourists flock to the coast. Long queues of traffic develop in the narrow lanes and the picturesque fishing villages heave with cars. The beaches become crowded and this can lead to environmental damage, with the erosion of paths over dunes and beautiful areas despoiled by thoughtless littering. In response to the influx of tourists, the government has commissioned a massive road building programme in the south-west which is the subject of heated debate. You can help minimise the damage by using public transport, or by visiting the area in the spring or autumn. Remember that out of season you can have huge expanses of golden sand to yourself.

About 40 kilometres to the south-west of Land's End, in the track of the North Atlantic Drift, lie the Isles of Scilly. This most westerly land of Great Britain comprises an archipelago of some 200 granite island and rocks separated from each other and the mainland by a shallow sea. The climate is mild with little variation in summer and winter temperatures. Many of the islands are small and devoid of vegetation, but the five largest – St Mary's, St Martin's, Bryher, Tresco and St Agnes – are inhabited and support farming, fishing and tourism industries; they are marvellous places to explore. Visitors to the Isles of Scilly are rewarded by beautiful scenery, shallow seas and a fascinating insight into the islands' past. Since no bathing water quality information is available to us we have not featured any individual beaches on the islands.

# WOOLACOMBE VILLAGE BEACH
## Devon
*OS Ref: SS456437 (see map opposite)*

Woolacombe village lies at the northern end of Woolacombe Sand, a two-mile stretch of glorious golden beach backed by sandy hills.

**Water Quality**
One outlet serving 13,200 people discharges secondary treated effluent 100 metres below LWM.

**Bathing Saftey**
Usually safe except at low tide and near the rocks. Emergency facilities and professional lifeguard cover between 10am and 6pm seven days a week from Whitsun to mid-September.

**Litter**
Cleaned daily from May to September; large bins are provided for litter. Dogs are banned from a zoned area of the beach during the season.

**Access**
Various paths, steps and slipways lead from the road and car park to the beach. Disabled visitors can reach the beach without having to use the steps, although the soft sand may cause difficulty for wheelchairs.

**Parking**
Three large car parks can accommodate nearly 3,000 vehicles. There is further free parking on the Esplanade and space in the village for 100 cars.

**Public Transport**
By National Express coach to Barnstaple and Ilfracombe or by rail to Barnstaple. Regular bus services run from Barnstaple and Ilfracombe.

**Toilets**
Full toilet facilities including baby changing and facilities for the disabled.

**Food**
Available from nearby cafés.

**Seaside Activities**
Swimming, surfing, windsurfing, donkey rides, swingboats, train ride. Speed boats and jet skis prohibited.

**Wet Weather Alternatives**
Mortehoe Heritage Centre, Once Upon A Time at the old Mortehoe Railway Station, Ilfracombe Museum, and other local attractions.

**Wildlife and Walks**
Woolacombe sands is within the Heritage Coastline and is backed by National Trust land and Morte Point to the north, which is also owned by the National Trust. These areas provide miles of spectacular coastal walks.

**Tourist Information**
Woolacombe Tourist Information Centre, Beach Road, Woolacombe, North Devon. Tel. 01271 870553

**Track Record GGPG**
EU designated beach.

# PUTSBOROUGH BEACH
### Devon
*OS Ref: SS447408*

This north-west facing beach at the southern end of Morte Bay provides superb bathing on the Atlantic coast, and is very popular with surfers the year round. The beach is reached by a narrow lane from Croyde.

 **Water Quality**
No routine sewage discharge has been identified.

 **Bathing Safety**
Generally safe; no lifeguard cover.

 **Litter**
Cleaned regularly, with litter bins provided. Dogs are banned from a zoned area from Easter to October.

 **Access**
Putsborough is well signposted from Croyde and Georgeham. Both steps and slopes lead to the beach. Access is suitable for disabled visitors.

 **Parking**
There is ample parking.

 **Toilets**
Toilets are available.

**Food**
A café sells light refreshments.

 **Seaside Activities**
Swimming and all non-powered water sports. Surfboard hire available.

 **Wet Weather Alternatives**
See Woolacombe opposite.

**Wildlife and Walks**
See Woolacombe opposite.

*The spectacular sweep of Woolacombe Sand viewed from the Down.*

 **Tourist Information**
See Woolacombe opposite.

**Track Record GGFG**
EU designated beach.

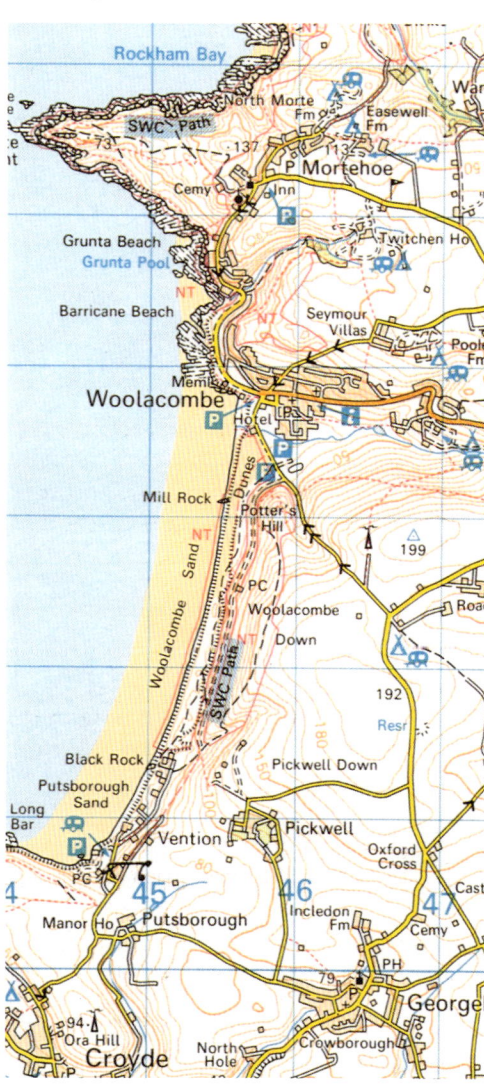

# HARTLAND QUAY
## Devon
*OS Ref: SS223248*

Dark jagged cliffs slide into the sea at Hartland Quay, whose small harbour was built by Raleigh, Drake and Hawkins as a safe haven for boats on this hazardous stretch of coast. The only buildings are a hotel converted from coastguard cottages, and a museum devoted to seafaring history and local wrecks, some of which can be seen at low tide. The beach is approached by a slipway built to replace the ancient harbour wall which was destroyed by storms at the turn of the century. A backdrop of spectacular cliff scenery gives way to pebbles and rocks, with sand exposed by the retreating tide.

### Water Quality
No routine sewage discharge has been identified.

### Bathing Safety
Safe within the Quay Bay; no lifeguard cover is provided and swimmers should take note of the sea conditions.

### Litter
Dogs are permitted on the beach at all times.

### Access
Approximately half a mile from Stoke village a private access leads to Hartland Quay, winding its way down to the Hotel, with spectacular views of the coastline.

### Parking
Three car parks give a total capacity of 300 cars.

### Toilets
There are toilet facilities within reach of the beach.

### Food
Refreshments are available from the hotel.

### Seaside Activities
Beach games, swimming. Certain sea conditions give rise to excellent surf, but due to the nature of the beach this is recommended for experienced surfers only.

### Wet Weather Alternatives
The nearby hamlet of Stoke has a

**Right:** *Sheer granite cliffs plunge hundreds of feet to the sea at Hartland Quay.*

*South-West England*

14th century church, St Nectan's, whose 130ft tower was built as a landmark for sailors. In the churchyard is Stranger's Hill, where some of the victims of local shipwrecks were buried. Nearby Hartland Abbey, a family house since the 16th century, contains Victorian and Edwardian photographs, and has a woodland walk through the Gardens to the coast. The house is open to the public on Wednesday and Sunday afternoons in summer. It was extensively rebuilt in the 18th century on the site of a monastery founded soon after 1157.

### Wildlife and Walks
The South West Coast Path follows high cliffs southwards to Speke's Mill Mouth, where a waterfall tumbles in stages down to a pebble beach. Three miles further on at Marsland Mouth a stream marks the border between Devon and Cornwall.

To the north, there are pleasant walks to Hartland Point. For those keen to experience six miles of glorious rocky coastline accessible only by foot, the Coast path continues east to the picturesque and historic village of Clovelly, where cars are banned and life seems to proceed much as it did one hundred years ago. Walkers should keep to the marked paths as the unstable cliffs can be dangerous. At the Milky Way, one and a half miles south of Clovelly, visitors can see birds of prey and falconry displays, as well as farm animals and sheep dogs.

### Tourist Information
Bude Visitors Centre, Cresent Car Park, Bude, Cornwall. Tel. 01288 354240

**Track Record GGGG**
EU designated beach.

# BUDE – SANDY MOUTH and CROOKLETS
## Cornwall
*OS Refs: SS202099, SS203072*

Backed by typical urban developments, with beach huts, shops and extensive parking facilities, these two distinct beaches join at low tide to form one vast stretch of sand. Adjacent to the beaches are grassy downs, much of which is owned by the National Trust.

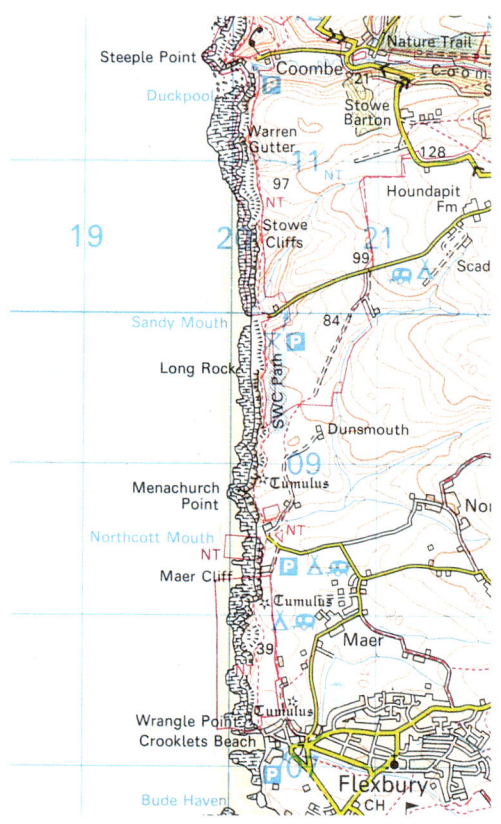

### Water Quality
No routine sewage discharge has been identified.

### Bathing Safety
Swimming is safe but be aware of rip currents, particularly at low tide. Lifeguards cover these beaches from May to early September.

### Litter
The beaches are cleaned all year round, daily in the summer. Dogs are banned from Easter to September.

### Access
From Bude, access to Sandy Mouth is via some steep steps after a loose gravel path. A concrete ramp at Crooklets allows disabled access, with a set of steps further along the beach.

### Parking
Capacity nearby for over 200 cars.

### Toilets
These include facilities for disabled visitors.

### Food
At Sandy Mouth there is a beach café owned by the National Trust. At Crooklets a café and beach shops serve light refreshments. Restaurants can be found in the town centre.

### Wet Weather Alternatives
There are plenty of wet weather facilities in and around Bude.

### Seaside Activities
Swimming and surfing.

### Wildlife and Walks
The North Cornwall Heritage Coast and Countryside Service operates a programme of guided walks throughout the district during the summer. There are specially devised circular walks in the area and the National Trust publish a leaflet on the Duckpool / Sandy Mouth properties. Free information leaflets on wildlife habitats, walking, cycling and activities within

*South-West England*

the area are available from the Bude Visitors Centre. The Tamar lakes to the north-east of Bude offer fishing and boats for hire.

**ℹ Tourist Information**
Bude Visitors Centre, Cresent Car Park, Bude. Tel. 01288 354240

*Sculpted by wind and weather, jagged pinnacles of rock rise above the sands like primitive pyramids.*

**Track Record**
**Sandy Mouth PGGG**
**Crooklets PPGG**
Both beaches are EU designated.

# WIDEMOUTH BAY
## Cornwall
*OS Ref: SS198024*

This large sandy beach with its adjacent rocky reefs provides good rockpooling opportunities. Backed by low cliffs and surrounded by fascinating rock formations, the bay is known for excellent surf.

### Water Quality
No routine sewage discharge has been identified.

### Bathing Safety
Generally safe; lifeguard cover from the end of May to September.

### Litter
The beach is cleaned regularly throughout the year and operates a zoned dog ban from Easter to October.

### Access
Widemouth can be found on the A39 south from Bude. The beach is accessible by steps and concrete ramp.

### Parking
Two large car parks provide space for over 300 cars.

### Public Transport
A bus service runs from Bude to Widemouth Bay.

### Toilets
These include facilities for disabled visitors.

### Food
Cafés and beach shops nearby.

### Seaside Activities
Swimming, rockpooling, surfing.

### Wet Weather Alternatives
Bude is not far away and has numerous facilities for rainy days.

### Wildlife and Walks
The coast path runs by the beach.

### Tourist Information
Bude Visitors Centre, Cresent Car Park, Bude. Tel. 01288 354240

### Track Record GPGG
EU designated beach.

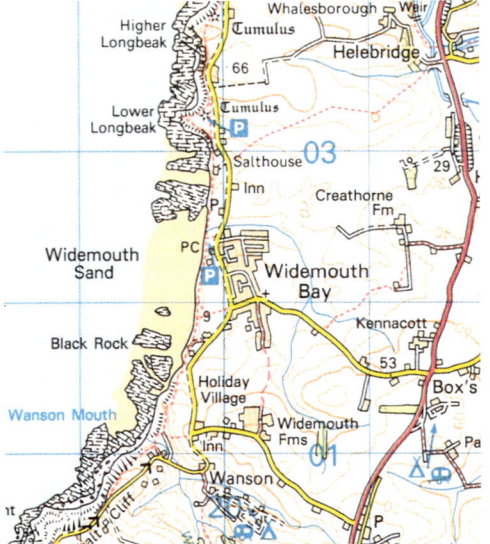

# DAYMER BAY
## Cornwall
*OS Ref: SW928776*

This wide sandy beach on the Camel Estuary is backed by sand dunes and a golf course. The water is shallow and ideal for bathing.

**Water Quality**
No routine sewage discharge has been identified.

**Bathing Safety**
Swimming is safe.

**Litter**
The beach is cleaned by the Parish Council.

**Access**
Take the Wadebridge to Polzeath road and turn left to Rock, then right for Daymer. Access to the beach is good but requires the use of some steps.

**Parking**
Adjacent to the beach.

**Toilets**
These include facilities for disabled visitors.

**Food**
A beach café opens in the season.

**Seaside Activities**
Swimming and windsurfing.

**Wet Weather Alternatives**
Information on the many activities can be obtained from the information point at Wadebridge and Polzeath Tourist Information Centre.

**Wildlife and Walks**
There are plenty of fine walks and numerous published circular routes. The North Cornwall Heritage Coast and Countryside Service publishes information on wildlife, walking, cycling and other activities, available from Tourist Information Centres throughout the district.

**Tourist Information**
Polzeath Tourist Information Centre, Coronation Gardens, Polzeath. Tel. 01208 862488

**Track Record PGGG**
EU designated beach.

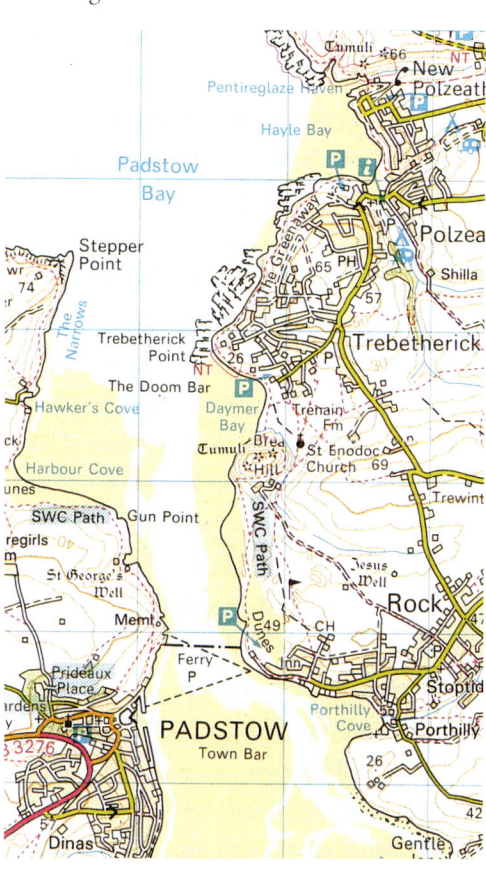

# CONSTANTINE BAY
## Cornwall
*OS Ref: SW857746 (see map opposite)*

This wide, sweeping arc of gently shelving pale sands, backed by large marram-covered dunes and bounded on either side by low headlands with rocky outcrops stretching seaward, is a picture to behold. Few facilities are available at the beach.

**Water Quality**
No routine sewage discharge has been identified.

**Bathing Safety**
Bathing is dangerous near the rocks. Lifeguards patrol from Whitsun to August Bank Holiday.

**Litter**
The beach is cleaned year round.

**Access**
Off the B3276 at St Merryn.

**Parking**
Capacity for 200 cars off the B3276 two minutes from the beach.

**Public Transport**
Nearest railway station is Bodmin Parkway; there are infrequent buses to Constantine Bay.

**Toilets**
At the entrance to the beach.

**Food**
There are shops ten minutes' walk from the beach.

**Seaside Activities**
Swimming and surfing.

**Wildlife and Walks**
See Treyarnon Bay opposite.

**Tourist Information**
See Treyarnon Bay opposite.

**Track Record GGGG**
EU designated beach.

# TREYARNON BAY
## Cornwall
*OS Ref: SW857740*

A wide sandy bay in an Area of Outstanding Natural Beauty, Treyarnon is sometimes overlooked by virtue of its location next to the larger Constantine Bay. Swimming is dangerous near the cliffs and surfing hazardous at low tide due to exposed rocks.

### Water Quality
No routine sewage discharge has been identified.

### Bathing Safety
Swimmers should keep to the centre of the beach where lifeguards patrol from Whitsun to August Bank Holiday. A natural pool in the rocks provides safe swimming at low tide.

### Litter
The beach is cleaned regularly.

### Access
Turn off the B3276 through Treyarnon. The Bedruthen Steps down to the beach have been restored by the National Trust.

### Parking
There is a small car park right next to the beach with approximately 150 spaces.

### Toilets
Toilets near the beach include facilities for disabled visitors.

### Food
Beach shop and hotel.

### Seaside Activities
Swimming and surfing; boards are available for hire.

### Wildlife and Walks
Treyarnon is on the North Cornwall Coast Path with spectacular views to Trevose Head in the north and Newquay to the south. Nearer the shoreline many rockpools are exposed at low tide providing a fascinating close-up of marine life.

### Tourist Information
Padstow Tourist Information Centre, The Old Red Brick Building, North Quay, Padstow. Tel. 01841 533449

### Track Record GGGG
EU designated beach.

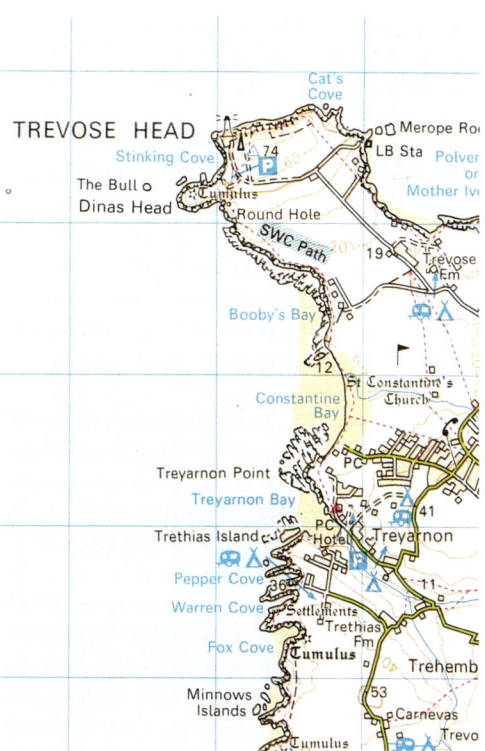

*Waves crash on to the rocks at Treyarnon Point, with the beach in the background.*

## PERRANPORTH – PENHALE SANDS and VILLAGE END BEACH
### Cornwall
*OS Refs: SW762570, SW757548*

Both beaches face west and form part of a three mile stretch of fine golden sand, surrounded by high dunes and cliffs. This is a Site of Special Scientific Interest.

### Water Quality
Penhale Sands – No routine sewage discharge has been identified.
Village End Beach – One outlet serving 12,000 people discharges disinfected secondary effluent at LWM.

### Bathing Safety
Lifeguard cover is provided but only at Village End Beach during the season. Bathing is allowed between the yellow and red flags.

### Litter
Village End Beach is cleaned daily; Penhale Sands weekly. Dogs are permitted on both beaches with a poop-scoop scheme in operation.

### Access
Penhale Sands is accessible through Haven Warner Holiday Park, off the coast road to Newquay; a footpath leads through the dunes from the car park. Village End Beach is accessed from Perranporth, where

South-West England

steps lead from the car park down to the beach.

### Parking
There are twenty spaces at Penhale Sands and 200 at Village End.

### Public Transport
Buses run daily to Perranporth from Truro and Newquay.

### Toilets
Toilets, at Village End only, include disabled and baby changing facilities.

### Food
There is a licensed café and restaurant at Village End.

### Seaside Activities
Swimming, surfing, and beach games.

### Wet Weather Alternatives
Perranporth has a Folk Museum and of course the local shops. Nearby St Agnes has craft workshops, a leisure park and museum.

### Wildlife and Walks
Details of walks and trails are available from the Tourist Centre.

### Tourist Information
Perranporth Tourist Information Centre, Perranporth, Cornwall, TR6 0DP. Tel. 01872 573368

**Track Record**
**Penhale Sands GGGG**
**Village End Beach PPPG**
Both beaches are EU designated.

*The retreating tide at Perranporth reveals a vast expanse of sand which is perfect for paddling and other beach games.*

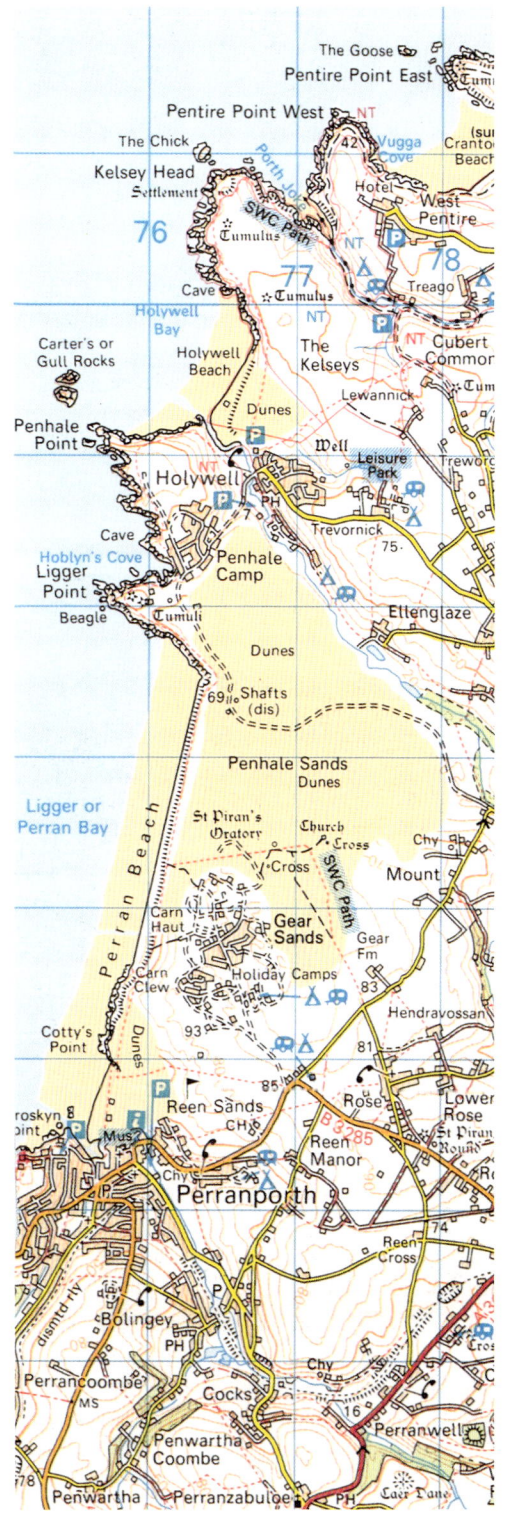

27

# ST IVES BAY
## Cornwall

This necklace of golden beaches is bordered to the west by the picturesque fishing town of St Ives and backed by the dunes of Hayle. Whether you enjoy swimming, surfing, fishing or simply exploring, St Ives Bay has something to offer. Five beaches within this 6-kilometre stretch of coast have excellent water quality, free from the problems of sewage contamination prevalent at a number of other Cornish beaches. They are:

Hayle – The Towans, OS Ref: SW563395
Carbis Bay – Porth Kidney Sands, OS Ref: SW540385
St Ives – Porthminster, OS Ref: SW522402
St Ives – Porthgwidden, OS Ref: SW522411
St Ives – Porthmeor, OS Ref: SW515410

### Water Quality
No routine sewage discharges have been identified at any of the listed beaches.

### Bathing Safety
Bathing is generally safe; follow the advice of lifeguards who patrol Hayle, Porth Kidney Sands, Porthminster and Porthmeor from May until September. Do not bathe in the estuary separating Hayle and Porth Kidney as strong currents here can make the waters hazardous.

### Litter
All beaches are cleaned regularly and most provide litter bins. Dogs are banned from Easter until October at all except Porth Kidney Sands.

### Access
Beaches to the west of the river are reached from the A3074, and to the east from the B3301. Steps or paths lead from adjacent roads. Access to Porth Kidney involves a short walk across the golf course.

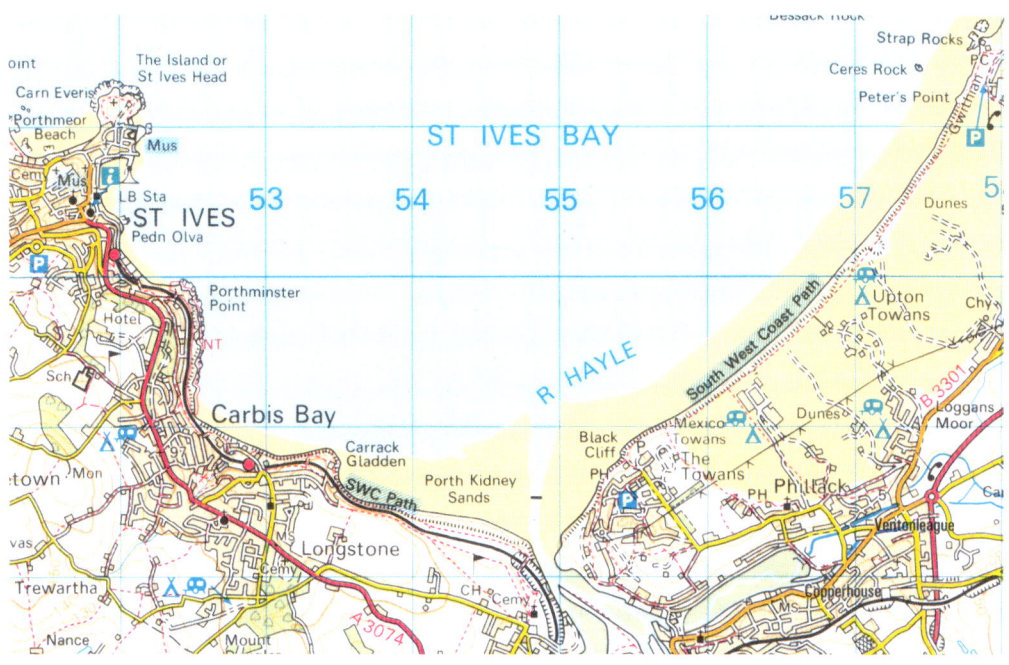

*South-West England*

### Parking
Facilities are located conveniently for all beaches. The car park serving Porth Kidney is relatively small.

### Public Transport
Nearest rail stations are Hayle, Lelant, Carbis Bay and St Ives. Regular bus services stop to set down and pick up close to all beaches.

### Toilets
All except Porth Kidney Sands have toilet facilities. Disabled toilets are located at Station Beach in Carbis Bay and at Porthmeor, St Ives.

### Food
Cafés and kiosks by the beaches together with nearby shops and pubs serve a wide range of hot and cold food and drink. Porth Kidney Sands, by contrast, is a rural beach and has only limited amenities.

### Seaside Activities
Swimming, surfing, windsurfing, snorkelling, canoeing, fishing.

### Wet Weather Alternatives
Bird of Paradise Park, Hayle; The Tate Gallery, St Ives; The Lifeboat House at St Ives. The towns of St Ives and Hayle

*The river Hayle separates the Towans from Porth Kidney Sands, with St Ives in the background.*

also offer a range of shops and other amusements, as well as a cinema.

### Wildlife and Walks
The beaches are served by the South West Coast Path which leads to interesting destinations in both directions, including to the west the village of Zennor with its 12th century church, and to the east, Godrevy Point with views across to Godrevy Island.

### Tourist Information
St Ives Tourist Information Centre, The Recreation Ground, Lethlean Lane, St Ives, Cornwall. Tel. 01736 796297

Hayle Tourist Information Centre, The Guildhall, Street-an-Pol, Hayle, Cornwall. Tel. 01736 754399

**Track Records**
**The Towans GGGG**
**Porth Kidney GPGG**
**Porthminster PFPG**
**Porthgwidden FFFG**
**Porthmeor GPGG**
All beaches are EU designated.

# PRIEST'S COVE, ST JUST
## Cornwall
*OS Ref: SW352317*

This attractive beach for fishing and relaxing has limited access to the sea, except via the slipway, but a children's swimming pool in the rocks more than compensates.

**Water Quality**
No routine sewage discharge has been identified.

**Bathing Safety**
Static lifesaving equipment and an emergency phone are provided on site.

**Access**
Via a footpath around the coast from St Just.

**Parking**
Car park at nearby St Just.

**Toilets**
Toilets are available.

**Seaside Activities**
Swimming and fishing.

**Wet Weather Alternatives**
The world famous Minack open-air theatre and St Michaels Mount are within easy driving distance.

**Wildlife and Walks**
There are many coastal and inland walks and trails. Penwith - Kerrier Ramblers (Tel. 01736 752121) welcomes visitors; programme details are available from the Tourist Information Centre.

**Tourist Information**
St Just Tourist Information Centre (Information Only), St Just Library, Market Square, St Just.
Tel. 01736 788669

**Track Record P~~G**
Not EU designated.

*South-West England*

# PRAA SANDS EAST, ASHTON, HELSTON
## Cornwall
*OS Ref: SW585276*

This attractive mile-long sweep of beach between two rocky headlands is sheltered by Hoe Point Cliffs to the west, edged by high dunes, and stretches east to Lesceave Cliff and the granite Rinsey Headland. Around 100 metres of gently sloping sand is exposed at low tide with rockpools at either end. Caravan parks nearby mean that the beach is often busy.

### Water Quality
No routine sewage discharge has been identified.

### Bathing Safety
Rip currents can develop at low tide; observe the warning signs. Lifeguards are on duty from May to September.

### Litter
Cleaned regularly during the summer season. Dogs are banned from Easter to the end of September.

### Access
Off the A394, between Penzance and Helston. The beach lies a short walk from the car park.

### Parking
There are two car parks close by, one on the road to the beach, the second at the bottom of the hill.

### Public Transport
Hourly bus services to Praa Sands from Penzance station.

### Toilets
At the entrance to the beach.

### Food
There are cafés, take-aways and shops.

### Seaside Activities
Swimming, diving, surfing, wind-surfing, fishing and sailing.

### Wildlife and Walks
The coast path from the eastern end of the beach leads up the cliff and on to Rinsey Head. Wheal Prosper, the engine house of an old copper mine, stands on the headland; the property and surrounding land is owned by the National Trust. A large reedbed five kilometres east of Penzance gives superb viewing of the rare Cetti's Warbler.

### Tourist Information
Helston Tourist Information Centre, 79 Meneage Street, Helston, Cornwall, TR13 8RB. Tel. 01326 565431

**Track Record GGGG**
EU designated beach.

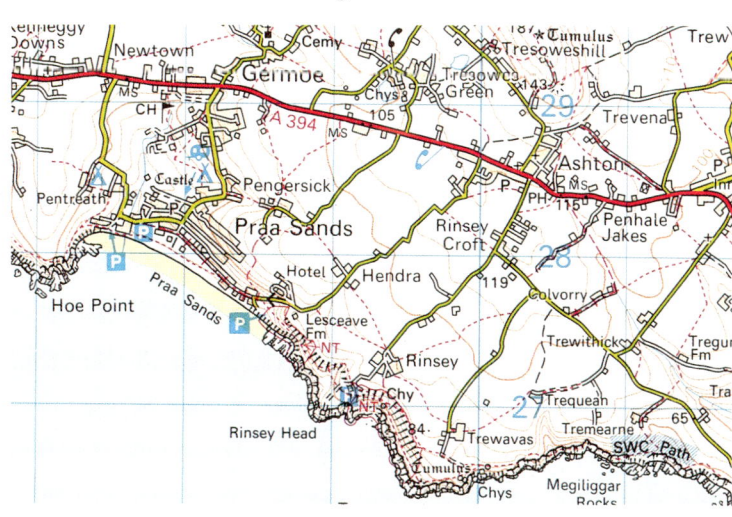

31

# PORTHPEAN
## Cornwall
*OS Ref: SX032507*

Once known for pilchard fishing, Porthpean is now popular amongst families seeking a safe, sheltered sandy beach. The cliffs to the east are extensively used by nesting birds, while the rockpools to the west provide ample opportunity to investigate marine life.

**Water Quality**
No routine sewage discharge has been identified.

**Bathing Safety**
Though not attended by lifeguards, bathing is considered very safe.

**Litter**
Cleaned daily during the summer.

**Access**
From the A390 follow the signs for Porthpean. Access to the beach is via a steep ramp leading from the car park, or by way of the coastal path.

**Parking**
There is pay car park 100 metres from the beach.

**Toilets**
Well maintained toilets are on site.

**Food**
Refreshments are available from a snack bar situated on the promenade.

**Seaside Activities**
Swimming, windsurfing; the beach is also home to Porthpean sailing club. Swimming and boating areas are segregated and jet-skiing is not allowed.

**Wet Weather Alternatives**
The bustling market town of St Austell is 2 miles away; its facilities include a cinema and leisure centre. Beer lovers might enjoy a guided tour of the 19th century St Austell Brewery.

*Right: Wooded cliffs and glistening sands give Porthpean a Continental feel.*

*South-West England*

### Wildlife and Walks
The South West Coast Path provides an easy route east to the Georgian port of Charlestown, named after Charles Rashleigh, a local mine owner who built the harbour in the late 18th century. A more vigorous walk to the west leads to Black Head, at the western limit of St Austell Bay. Nesting birds occupy the cliffs to the east and rock pools are located at the west end of the beach.

### Tourist Information
Porthpean Tourist Information Centre, Bypass Garage, Southbourne Road, St Austell, Cornwall.
Tel. 01726 76333

**Track Record GGGG**
EU designated beach.

# CHALLABOROUGH
## Devon
*OS Ref: SX649448 (see map opposite)*

This sheltered horseshoe-shaped cove with sand, fine shingle and rocks is backed by the holiday village of Challabrough and has extensive rock pools to explore at low tide.

 **Water Quality**
No routine sewage discharge has been identified.

 **Bathing Safety**
Safe with normal precautions. Lifeguards present between May and September.

 **Litter**
The beach is cleaned five times a week by hand and twice a week by tractor. Dogs are banned between May and September.

*Fine sand makes Challaborough a good sandcastle beach.*

**Access**
Take the B3392 off the A379 from Kingsbridge to Plymouth. The beach is off a minor road beyond Ringmore.

**Parking**
Adjacent to the beach.

**Public Transport**
Bus service 87, Monday to Saturday from Ivybridge to Bigbury, one kilometre away.

 **Toilets**
Toilets are in the car park.

 **Food**
Fish and chips are available, and pubs serving bar food.

**Seaside Activities**
Swimming, surfing and walking.

**Wet Weather Alternatives**
There is a sports and leisure club next to the beach.

**Wildlife and Walks**
A fine 4-mile walk on the South West Coast Path along high wild cliffs leads north-west to Wonwell.

 **Tourist Information**
Kingsbridge Tourist Information Centre, The Quay, Kingsbridge. Tel. 01548 853195

**Track Record GGGG**
EU designated beach.

# THURLESTONE NORTH and SOUTH MILTON SANDS
## Devon
### OS Ref: SX674421, SX676417

These two beaches to the south of the village of Thurlestone in an Area of Outstanding Natural Beauty offer coarse sand with rocky outcrops and plenty of rockpools to explore at low tide. The most prominent feature in the area is the famous Thurlestone Rock, painted by Turner and best seen at high tide.

### Water Quality
No routine sewage discharge has been identified at either beach.

### Bathing Safety
Safe, but follow advice given by lifeguards and observe warning flags.

### Litter
Dogs are banned from both beaches, which are cleaned regularly during the summer.

### Access
Thurlestone is on minor roads off the A379 Kingsbridge to Modbury road.

### Parking
Three car parks serve the two beaches.

### Toilets
Toilets are situated at the northermost and southernmost car parks.

### Food
There is a hotel and a beach café at the southernmost of the three car parks.

### Seaside Activities
Swimming, surfing, windsurfing (there is a windsurfing school at the beach), snorkelling, golf and walking.

### Wildlife and Walks
There are clifftop walks from National Trust land at Bolberry Down, just two miles to the south. South Milton Ley is the second largest reedbed in Devon and an important habitat for migrating birds.

### Tourist Information
Kingsbridge Tourist Information Centre, The Quay, Kingsbridge, Devon. Tel. 01548 853195

**Track Record**
**Thurlestone North GGGG**
**South Milton Sands GGGG**
Both beaches are EU designated.

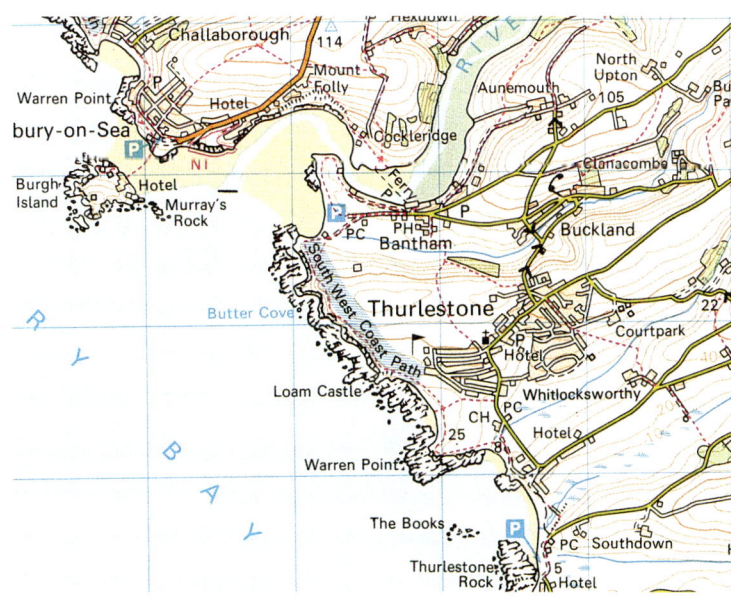

South-West England

# SALCOMBE NORTH SANDS
## Devon
*OS Ref: SX731382*

Situated on the southern outskirts of Salcombe, this is a small enclosed sandy beach at the mouth of the estuary with views across the harbour entrance.

**Water Quality**
No routine sewage discharge has been identified.

**Bathing Safety**
Bathing is generally safe and some basic emergency equipment is provided on site.

**Access**
Down a steep winding lane south of Salcombe.

**Parking**
A pay-and-display car park for 87 cars lies immediately above the beach.

**Toilets**
There are toilets within reach of the beach.

**Food**
A café is situated above the beach.

**Seaside Activities**
Swimming and windsurfing.

**Wet Weather Alternatives**
Salcombe Town and Harbour to the north; Overbecks House (National Trust) half a mile south.

**Wildlife and Walks**
There is a plesant footpath up the wooded valley behind the beach.

**Tourist Information**
See Mill Bay (opposite).

**Track Record PPPG**
EU designated beach.

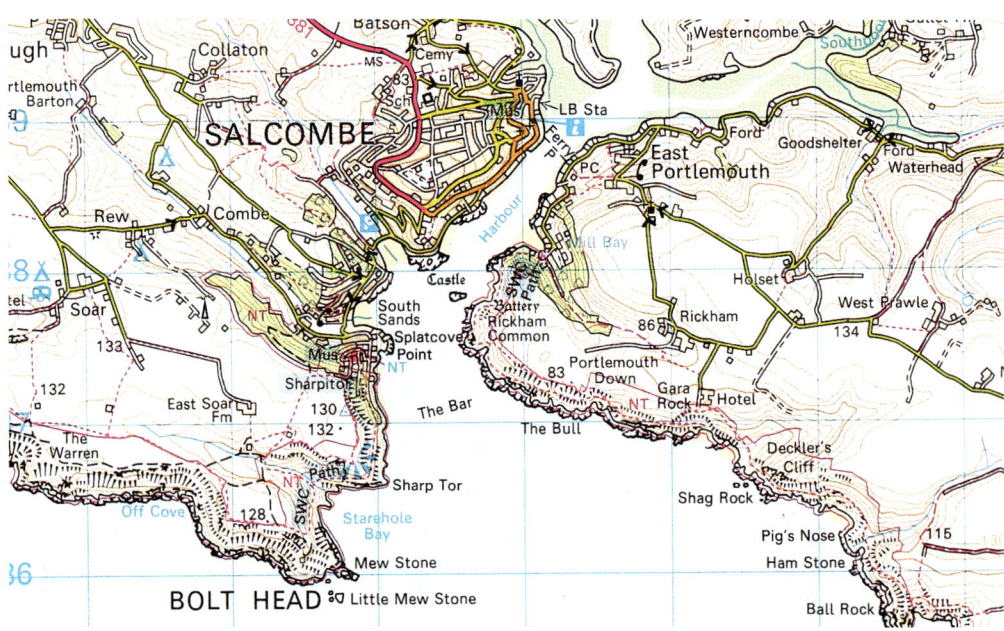

*South-West England*

# MILL BAY near SALCOMBE
## Devon
*OS Ref: SX740382 (see map opposite)*

A privately owned beach across the estuary from Salcombe, this fine sandy cove is sheltered and set in a beautiful rural area with gentle slopes down to the beach. The picturesque village of East Portlemouth is about a kilometre away although it is a steep climb to get there by foot. During the Second World War Mill Bay was used as an American base. Air-sea rescue boats were moored in the harbour and an anti-aircraft battery was sited on the cliffs above the beach.

**Water Quality**
No routine sewage discharge has been identified.

**Bathing Safety**
There is no lifeguard cover but bathing is generally safe.

**Litter**
The beach is cleaned regularly.

**Access**
The easiest approach to the beach is by passenger ferry from Salcombe.

**Parking**
There is a small car park next to the beach; note that both this and the narrow approach road become very congested in summer.

**Toilets**
Adjacent to the beach.

**Food**
The nearest café is 1km away.

**Seaside Activities**
Swimming, windsurfing, walking.

**Wet Weather Alternatives**
None at this beach.

**Wildlife and Walks**
Walk leaflets are available from the Tourist Information Centre. Prawle Point, the southernmost tip of Devon, (SX772350) is nearby. This is a rocky coastline with areas of steep scrub, open grassland and farmland, where the cirl bunting (a Mediterranean species in decline in Britain) breeds.

**Tourist Information**
Salcombe Tourist Information Centre, Market Street, Salcombe, TQ8 8DE. Tel. 01548 843927

**Track Record GGGG**
EU designated beach.

*Just far enough off the beaten track to keep the hordes away, Mill Bay is still popular with visitors in the summer months.*

# BEESANDS
## Devon
*OS Ref: SX820405*

This mile-long shingle beach in a very quiet rural location is backed by a large area of grassland and a freshwater lake. A sea wall protects the line of cottages and a pub that make up the village. The coast path runs nearby.

### Water Quality
No routine sewage discharge has been identified.

### Bathing Safety
Generally safe; basic lifesaving equipment is on site.

### Litter
The beach is cleaned daily during the summer, five days a week by hand and twice a week mechanically. There is no dog ban in operation.

### Access
Signposted from Stokenham, between Kingsbridge and Dartmouth.

### Parking
There is extensive free parking immediately beside the beach.

### Toilets
Toilet facilities are available.

### Food
A café and pub in Beesands village.

### Seaside Activities
Swimming.

### Wildlife and Walks
The lake behind the beach provides interesting birdwatching.

### Tourist Information
Kingsbridge Tourist Information Centre, The Quay, Kingsbridge. Tel. 01548 853195

### Track Record ~P~G
Not EU designated.

*South-West England*

# SLAPTON SANDS
## Devon
*OS Ref: SX828440 (see map opposite)*

Slapton Sands (a misleading name since it is predominantly shingle) is a three-mile long straight beach backed by the large freshwater lake of Slapton Ley, a National Nature Reserve. The villages of Strete Gate and Torcross are located at the northern and southern ends of the beach respectively.

### Water Quality
No routine sewage discharge has been identified.

### Bathing Safety
Generally safe; basic lifesaving equipment is provided at Torcross and Strete Gate.

### Litter
The beach is cleaned daily during the summer, five days a week by hand and twice a week mechanically. There is no dog ban in operation.

### Access
Slapton is situated alongside the A379 Kingsbridge to Dartmouth road.

### Parking
Three car parks, one at either end of the beach and one in the middle, have a total of 500 parking spaces.

### Public Transport
A regular service is provided by the number 93 bus which operates between Plymouth and Dartmouth.

### Toilets
Toilet are at Monument car park.

### Food
Refreshments can be obtained from Slapton or Torcross.

### Seaside Activities
Swimming.

### Wet Weather Alternatives
Half-way along the beach is a war memorial dedicated to the victims of Operation Tyger, a D-Day landing exercise that ended in tragedy when the craft were sunk by German E-boats. A restored Sherman tank recovered from the sea stands in Torcross.

### Wildlife and Walks
Slapton Ley Nature Trail lies behind the beach. For further information on walks contact the local Tourist Information Centre.

### Tourist Information
Dartmouth Tourist Information Centre, Newcomen Engine House, Mayors Avenue, Dartmouth. Tel. 01803 834224

**Track Record GGPG**
EU designated beach.

# BLACKPOOL SANDS, STOKE FLEMING
## Devon
*OS Ref: SX855478*

Blackpool Sands stands in complete contrast to its Lancashire cousin: the only development on this beach comprises a car park, a shop and attractive toilet block. This crescent shaped stretch of golden sands is surrounded by green fields, magnificent pines, craggy cliffs and a turquoise blue sea.

### Water Quality
No routine sewage discharge has been identified.

### Bathing Safety
Normally safe; lifebuoys are positioned along the beach and a rescue boat is available.

### Litter
The beach is cleaned daily and bins are located along its length. Dogs are banned from Easter to November.

### Access
Off the A379 between Stoke Fleming and Strete Gate; slipways provide access to the beach.

### Parking
There is ample parking.

### Public Transport
A regular service is provided by the number 93 bus which operates between Kingsbridge and Dartmouth.

### Toilets
Toilets are available, including facilities for the disabled and baby changing.

### Food
A kiosk and vending machines sell hot and cold food and drinks.

### Seaside Activities
Swimming; a water sports centre offers tuition on, and hire of, pico lasers, surf canoes, windsurfers, and an introductory sub-aqua course.

### Wet Weather Alternatives
Dartmouth Castle and Woodland Leisure Park.

### Wildlife and Walks
Inland and coastal walks, including a circular walk from the beach to Stoke Fleming that takes in the 13th-century church, the Green Dragon pub and dramatic coastal views.

### Tourist Information
See Slapton (previous page).

**Track Record GGGG**
EU designated beach.

*South-West England*

# SHOALSTONE BEACH
## Devon
*OS Ref: SX932566*

This shingle beach close to the old fishing port of Brixham affords excellent views across Torbay. Low tide exposes pools between angled tablets of rock. There is an open-air swimming pool at the eastern end of the beach.

### Water Quality
No routine sewage discharge has been identified.

### Bathing Safety
Trained beach attendants operate here.

### Litter
The beach is cleaned daily. Dogs are banned from May to September.

### Access
A left turn off the road from Brixham Harbour to Berry Head.

### Parking
Adjacent to the beach.

### Public Transport
A local bus service runs to Shoalstone.

*Shoalstone is well suited for those who prefer to just sit and admire the view.*

### Toilets
Adjacent to the beach.

### Food
There is a café near the beach.

### Seaside Activities
Swimming, diving, fishing and rock-pooling.

### Wet Weather Alternatives
Brixham Aquarium and museums, Berry Head Country Park.

### Wildlife and Walks
Shoalstone is situated close to Berry Head Site of Special Scientific Interest. There are numerous walks and trails in the area; contact the local Tourist Information Centre for details.

### Tourist Information
See Torre Abbey (overleaf).

**Track Record GGPG**
EU designated beach.

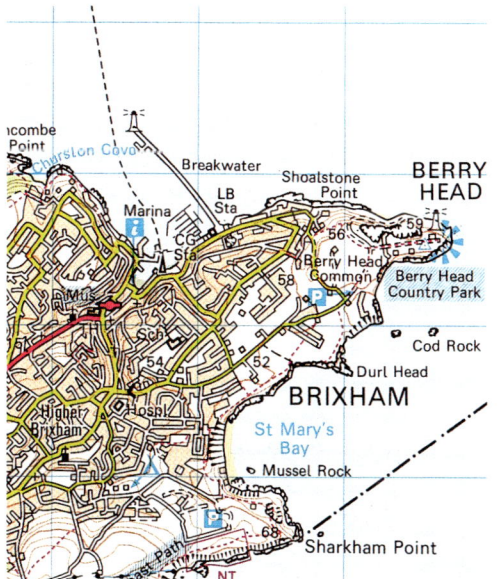

# TORRE ABBEY SANDS
## Devon
*OS Ref: SX909635 (see map opposite)*

This is a large sandy beach with a small rocky cove next to the headland. Rockpools are abundant and the beach is backed by a promenade and path.

### Water Quality
No routine sewage discharge has been identified.

### Bathing Safety
Generally safe; qualified beach attendants are present during the summer season.

### Litter
The beach is cleaned daily. Dogs are banned from May to September and a poop-scoop scheme is in operation.

### Access
On the main road from Torquay; follow signposts to the harbour.

### Parking
A nearby car park has space for 1,000 vehicles.

### Toilets
Toilets include facilities for disabled visitors.

### Food
There are cafés, restaurants and pubs nearby.

### Seaside Activities
Swimming, pedalos, floats and parascending.

### Wet Weather Alternatives
Torre Abbey. See also Meadfoot Beach (p44).

### Wildlife and Walks
The Agatha Christie Mile; the South West Coast Path and nearby parks and gardens.

*The gently sloping beach at Torre Abbey Sands makes for safe bathing in most weathers.*

### Tourist Information
Torquay Tourist Information Centre,
The Tourist Centre,
Vaughan Parade, Torquay.
Tel. 01803 297428

**Track Record PPPG**
EU designated beach.

*South-West England*

# BEACON COVE
## Devon
*OS Ref: SX919630*

This small pebbled cove adjacent to the harbour and close to the town centre provides some respite from the bustle of a busy resort.

**Water Quality**
No routine sewage discharge has been identified.

**Bathing Safety**
Generally safe; there is no lifeguard cover.

**Litter**
The beach is cleaned daily. Dogs are not banned, but a poop-scoop scheme is in operation.

**Access**
Follow the signposted route to the harbour car park; from here the beach is accessible via steps.

**Parking**
There is parking close by for over 500 cars.

**Toilets**
None in the vicinity.

**Food**
Refreshments are available in the town centre.

**Seaside Activities**
Swimming.

**Wet Weather Alternatives**
See Meadfoot Beach (overleaf).

**Wildlife and Walks**
See Torre Abbey (opposite).

**Tourist Information**
See Torre Abbey (opposite).

**Track Record PGGG**
EU designated beach.

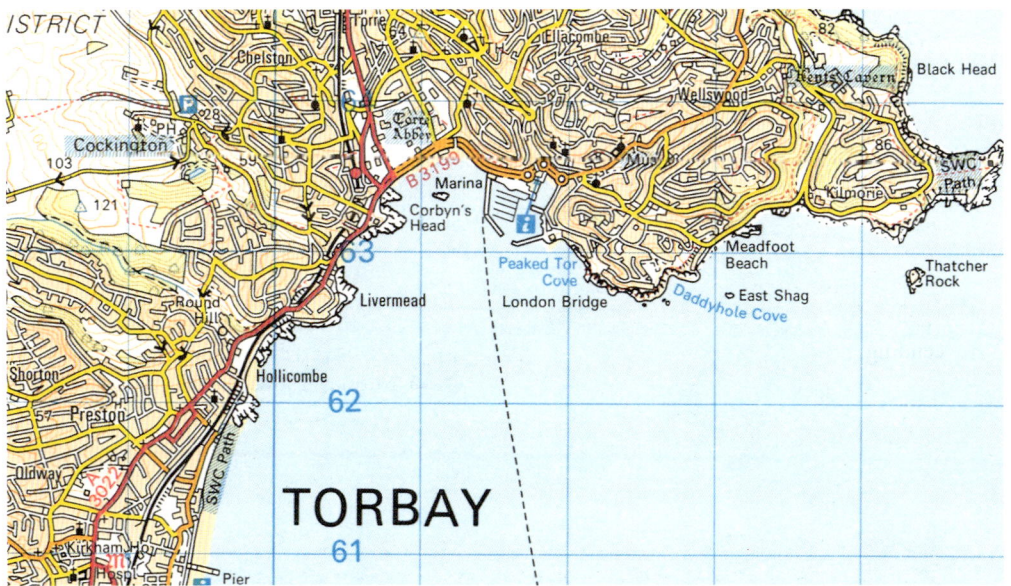

# MEADFOOT BEACH, TORQUAY
## Devon
*OS Ref: SX930630*

This sandy beach is situated just to the east of Torquay and south of Babbacombe Bay, in the middle of one of Britain's best known and most popular resort areas. The beach itself is backed by imposing cliffs which provide an ideal vantage point for observing the rest of the Tor Bay coastline. Should the occasional wet day be encountered, there are plenty of alternative attractions in the immediate vicinity.

### Water Quality
No routine sewage discharge has been identified.

### Bathing Safety
Generally safe; trained lifesavers are present during the summer.

### Litter
The beach is cleaned regularly and dogs banned from May to September.

### Access
From Torquay harbour, follow the signs for Babbacombe, turning right at the sign for Meadfoot Beach. The beach can be reached from the road.

### Parking
In car park adjacent to the beach.

### Public Transport
Torquay is the nearest rail station; a bus service runs to the beach.

### Toilets
Toilet facilities are available.

### Food
The beach has a café, and there are further shops within easy reach.

### Seaside Activities
Swimming, with pedal boats and floats available for hire.

### Wet Weather Alternatives
The Riviera Centre in Torquay, Paignton Zoo, Torbay Aircraft Museum, Kent's Cavern, Babbacombe Model Village, Berry Pomeroy castle and the Dart Valley

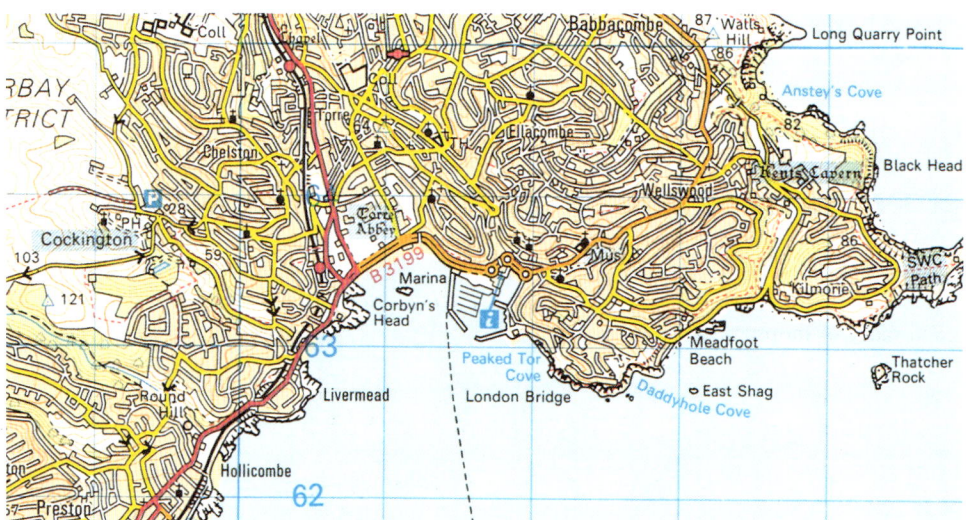

*South-West England*

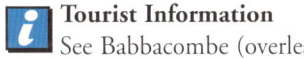

Railway are all within easy reach. There is a good range of shops and amusements in Torquay.

### Wildlife and Walks
The beach is adjacent to a Site of Special Scientific Interest. Meadfoot is also served by the South West Coast Path.

*Lazing away the day on Meadfoot Beach.*

### Tourist Information
See Babbacombe (overleaf).

**Track Record GPGG**
EU designated beach.

South-West England

# BABBACOMBE BAY, TORQUAY
## Devon

The stretch of coast surrounding the popular resort of Torquay benefits from a mild climate; with its abundant palm trees and other subtropical vegetation, it is not hard to see why it is known as the 'English Riviera'. Babbacombe Bay to the north of Torquay is particularly favoured, with six beaches in one 10-kilometre stretch which meet the highest standards of water purity. These are:

Anstey's Cove (Redgate Beach), OS Ref: SX930648
Babbacombe Beach, OS Ref: SX930654
Oddicombe Beach, OS Ref: SX926657
Watcombe Beach, OS Ref: SX926673
Maidencombe, OS Ref: SX927685
Ness Cove, OS Ref: SX938717

**Water Quality**
No routine sewage discharge has been identified at these beaches.

**Bathing Safety**
Generally safe, with trained beach attendants at Anstey's Cove and Oddicombe; swimmers should observe the safety flags at all times.

**Litter**
Beaches are cleaned daily over the summer season.

**Access**
All beaches are accessible off the main Torquay to Teignmouth road via steps or paths; these are steep at Anstey's Cove, Maidencombe and Watcombe. Oddicombe is served by a cliff railway.

**Parking**
Parking is widely available along this stretch of coast, though sometimes in the towns and at a little distance from the beaches.

*The dark red sandstone cliffs backing Oddicombe Beach stand out starkly against the golden sand.*

# South-West England

### Public Transport
Anstey's Cove, Oddicombe, Watcombe and Maidencombe are served by bus from Torquay, and Ness Cove from Teignmouth.

### Toilets
All the beaches have toilets and some include facilities for disabled visitors.

### Food
A wide range of food and drink is available from cafés, kiosks and nearby shops and pubs.

### Seaside Activities
Swimming, fishing, windsurfing, sailing, boat and pedalo hire.

### Wet Weather Alternatives
See Meadfoot (previous page).

### Wildlife and Walks
The South Devon Coast Path extends around the bay, offering some of the most glorious views on the south coast. The mild climate encourages unusually luxuriant plant life and the area is favoured by rare butterflies. From Ness Cove the Teign Estuary can easily be explored and the Wildlife Trust has a centre near the beach.

### Tourist Information
Torquay Tourist Information Centre,
The Tourist Centre,
Vaughan Parade, Torquay.
Tel. 01708 297428

**Track Record**
**Anstey's Cove (Redgate Beach) PPGG**
**Babbacombe Beach PPGG**
**Oddicombe Beach GGGG**
**Watcombe Beach PGGG**
**Maidencombe ~GGG**
**Ness Cove ~GGG**
All beaches are EU designated.

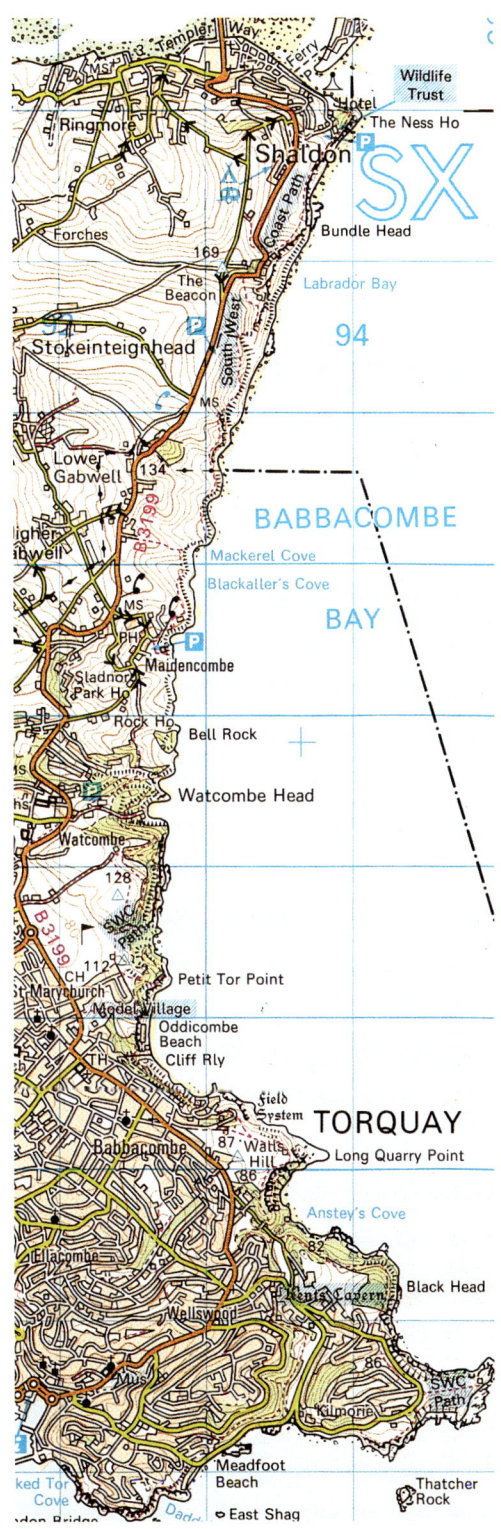

47

# EXMOUTH
## Devon
*OS Ref: SY009799*

This stylish and spacious resort boasts immaculate gardens and parks, some dating from the town's Victorian heyday, and some fine vernacular architecture from the period hints at a glorious past in more genteel times. From the old docks area there are magnificent views up river towards Exeter and in the summer a passenger ferry sails to the village of Starcross on the opposite bank of the Exe estuary. Exmouth's long sandy beach is backed by a sea wall and Queens Drive, a wide promenade lined with shops, restaurants and pubs. It is sheltered from all but the westerly winds.

### Water Quality
One discharge serving 42,700 people discharges secondary effluent 170 metres below LWM.

### Bathing Safety
Bathing is generally safe, but swimmers should beware of the strong tides. There is no lifeguard cover; unsafe bathing areas are marked by flags.

### Litter
The beach is cleaned daily. Dogs are banned from the beach between Easter and the end of September.

### Access
On entering the town, follow the signs to the seafront: there is direct access to the beach from Queens Drive which runs parallel to the shore.

### Parking
Numerous car parks along the promenade provide a combination of free and pay-and-display parking for over 1,000 vehicles.

### Toilets
Well maintained toilets include facilities for disabled visitors.

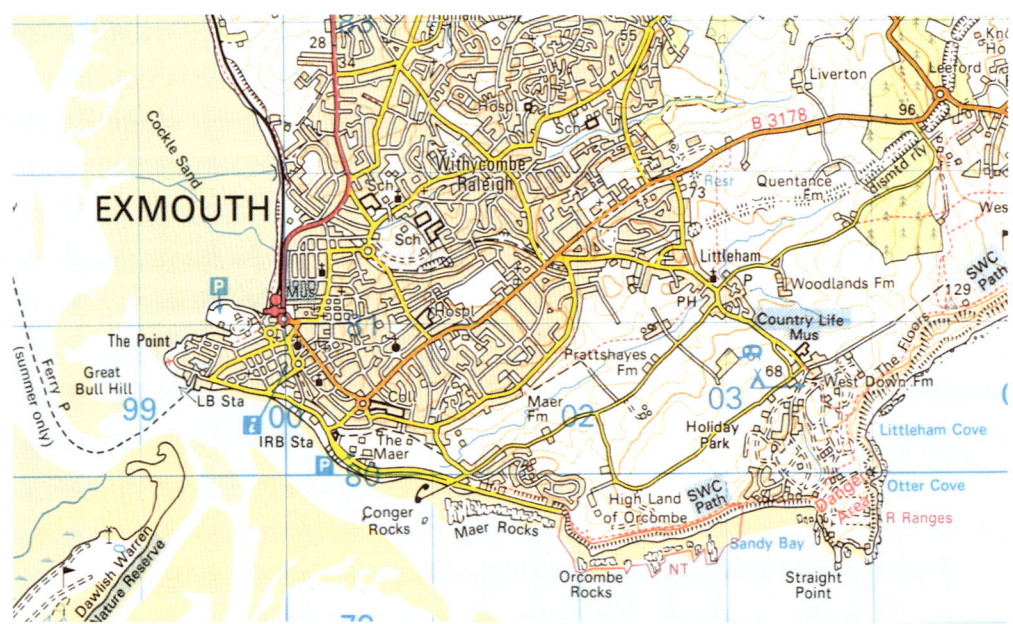

## South-West England

### Food
A wide range of refreshments are available, both in the town and from kiosks, pubs and restaurants on the promenade.

### Seaside Activities
Swimming, jet skiing, boat trips and windsurfing.

### Wet Weather Alternatives
Amusement arcades, aquarium and model railway. Just to the north of the town is A La Ronde, a National Trust property. This unique 16-sided house was built by its eccentric owner in 1796. A number of its rooms have walls encrusted with a mosaic of shells, feathers and other unusual items.

### Wildlife and Walks
The South Devon Coast Path and the East Devon Way run nearby. Exmouth Nature Reserve encompasses Cockle Sand and the surrounding mud flats. From the footpath or the drive-in path near the rail-

*The red sandstone cliffs are a particular feature of this section of the South Devon coastline.*

way station, visitors may see redshanks, dunlins and even sanderlings and grey plovers. Farther east an area of sand dunes and grassland forms Maer Nature Reserve, where around 400 species of plant grow. The internationally renowned nature reserve at Dawlish Warren, just across the estuary, also attracts large numbers of waders and wildfowl. Among its hundreds of flowering plants is the Warren crocus, whose lilac-blue flowers can be seen nowhere else in mainland Britain.

### Tourist Information
Exmouth Tourist Information Centre, Manor Gardens, Exmouth. Tel. 01395 222299

**Track Record FPGG**
EU designated beach.

# LYME REGIS - COBB/TOWN BEACH
## Dorset
*OS Ref: SY339918*

This pretty harbour town has an international reputation amongst palaeontologists, thanks to its cliffs which have yielded a vast array of fossils over many years. Latterly it has become more widely known as the scene of John Fowles' *The French Lieutenant's Woman*. The sandy beach featured here adjoins the Cobb – a large stone breakwater – made famous in the subsequent film version of the novel.

### Water Quality
One outfall serving 6,000 people discharges tertiary treated (UV disinfected) effluent 400 metres below LWM.

### Bathing Safety
Bathing is safe within the marked areas; lifesaving equipment is situated on the promenade.

### Litter
The beach is cleaned regularly and dogs are banned between April and September.

### Access
Once in the town, follow the signposts to the Cobb. The beach is accessible from the road.

### Parking
There are a number of local short-stay car parks.

### Toilets
Public toilets are situated in Broad Street and adjacent to the Cobb.

### Food
A range of outlets in the town provide food and refreshments.

### Seaside Activities
Swimming, windsurfing, fishing; boat trips and boat hire are available.

### Wet Weather Alternatives
Attractions include a museum housing a collection of locally excavated fossils,

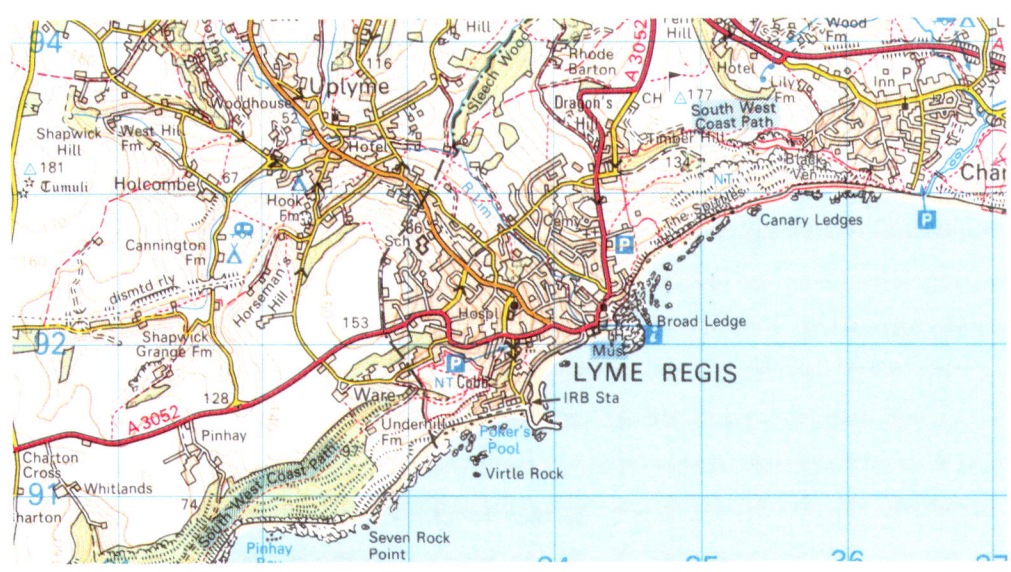

*South-West England*

an aquarium and a Heritage Coast Centre at nearby Charmouth

*Lyme Regis is as popular today as it was in Victorian times.*

### Wildlife and Walks

A coastal path guides visitors along the famous fossil-bearing cliffs. The area forms part of the West Dorset Heritage Coast, and rangers provide a guided walk service. This area of West Dorset is also designated an Area of Outstanding Natural Beauty.

### Tourist Information

Lyme Regis Tourist Information Centre, Church Street, Lyme Regis, DT7 3BS. Tel. 01297 442138

**Track Record FPPG**
EU designated beach.

# SEATOWN
## Dorset
*OS Ref: SY418916 (see map opposite)*

Reached by a lane from Chideock, this village with thatched cottages of honey-coloured stone has an open shingle beach which shelves steeply above sands exposed at low tide. The beach is backed by blue lias clay cliffs which are unsafe and dangerous to climb.

### Water Quality
One discharge serving 2,000 people discharges secondary treated effluent.

### Bathing Safety
Normally safe with the usual precautions. There is no lifeguard cover on this beach.

### Litter
Litter bins are located along the beach. Dogs are allowed and a poop-scoop scheme is in operation.

### Access
The beach is reached via Duck Street, Chideock, off the A35. Pebbles make access difficult for wheelchairs.

### Parking
There are spaces for around 200 cars.

### Public Transport
The Bridport to Lyme Regis bus stops at Chideock.

### Toilets
Toilets are available nearby.

### Food
The Anchor Inn beside the beach serves light meals and snacks.

### Seaside Activities
Swimming and fishing.

### Wet Weather Alternatives
Many local attractions include leisure centres at Bridport and Lyme Regis.

### Wildlife and Walks
The golden-orange sandstone peak of Golden Cap soars to over 600ft above sea level, making it the tallest cliff in southern England. Dramatic views from the top stretch as far as Portland Bill to the east and Start Point to the west. The cliff is part of a National Trust estate that embraces most of the coastal land between Eype and the Devon border, with approximately 30 miles of walks over terrain ranging from steep cliffs to undulating meadows and clumps of ancient woodland. Several paths lead to the cliff top: the shortest route is from the car park at Langdon Hill to the north-east.

### Tourist Information
Lyme Regis Tourist Information Centre, Church Street, Lyme Regis, DT7 3BS. Tel. 01297 442138

**Track record PPGG**
EU designated beach.

# EYPEMOUTH
## Dorset
*OS Ref: SY446910*

This small secluded shingle beach is backed by crumbling clay and sandstone cliffs which are unsafe and dangerous to climb. Some sand may be visible depending on the tides.

 **Water Quality**
No routine sewage discharge has been identified.

**Bathing Safety**
Safe; there is no lifeguard cover.

**Litter**
Litter bins are provided. Dogs are allowed, with a poop-scoop scheme in operation.

**Access**
Follow signs to Lower Eype from the Bridport bypass. A well maintained flight of steps leads from a pay car park to the beach.

**Parking**
There are spaces for approximately 75 vehicles in the car park.

 **Toilets**
Toilets are available.

**Food**
Hotels, pubs, and tea rooms serve light refreshments in Lower Eype.

**Seaside Activities**
Swimming and fishing.

**Wet Weather Alternatives**
See Seatown (opposite).

**Wildlife and Walks**
See Seatown (opposite).

 **Tourist Information**
See Seatown (opposite).

**Track record GPGG**
EU designated beach.

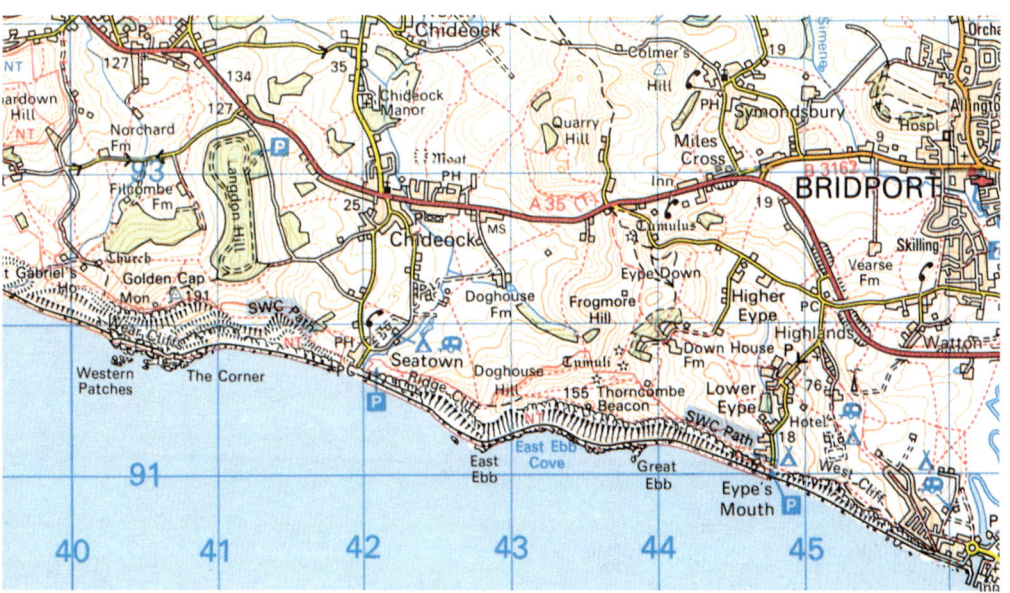

# WEYMOUTH CENTRAL and LODMORE
## Dorset
*OS Refs: SY681794, SY688807*

About 1.5 kilometres in length and ranging from 40 to 150 metres in width on mean tides, the beach at Weymouth is mainly of fine sand with shingle and pebbles at the north end. The bay is sheltered, with no dangerous tidal drifts or undercurrents, and renowned for its safety; swimmers, bathers and other water-users benefit from the gradual slope of the seabed which gives shallow inshore waters. Just to the north, and adjacent to a 300-acre site incorporating an internationally acclaimed RSPB Nature Reserve and Country Park, the shingle and pebble beach at Lodmore is perfect for a safe and enjoyable family day out.

### Water Quality
No routine sewage discharge has been identified at either beach.

### Bathing Safety
The inshore water has a speed limit of eight knots, and water sports and bathing areas are segregated from each other. Council beach control staff enforce the water safety bylaws, assisted by the harbour authority.

### Litter
Both beaches are cleaned daily by hand and machine; bins are provided at strategic points. Dogs are banned from the beaches and other designated areas between May and September. A poop-scoop scheme is in operation.

### Access
Weymouth is signposted from the M27 (east), M5 (west) and A38 (south). Lodmoor can be reached from the A353 to the north.

### Parking
Car parks in Weymouth provide

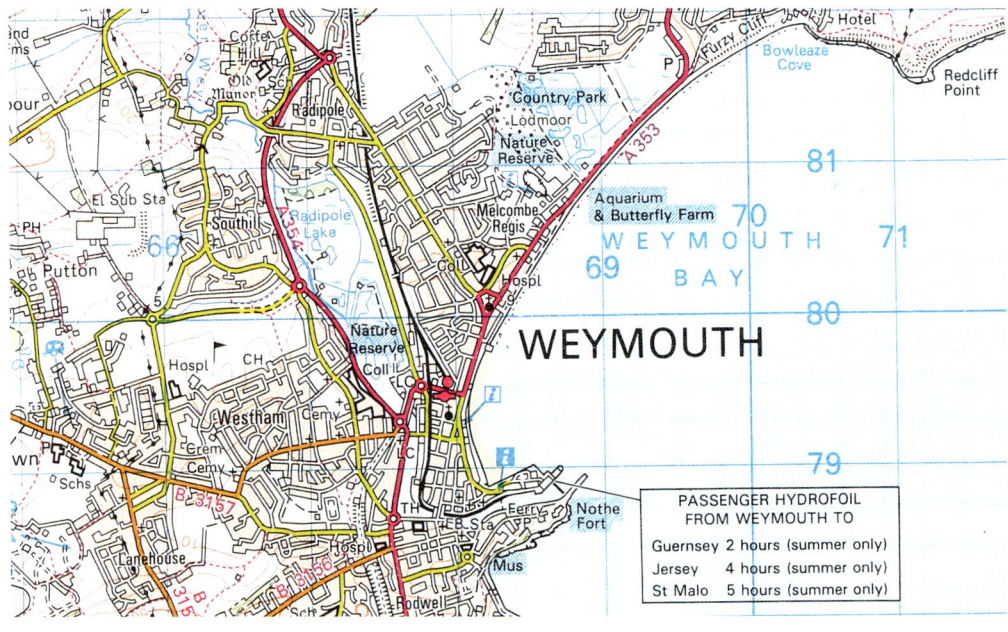

# South-West England

*Weymouth's sandy beach is accessible all along its length from the road.*

ample capacity. For Lodmoor, use the Country Park and beach car parks.

### Public Transport
Weymouth is served by rail, by National Express coaches and by frequent buses from many local towns.

### Toilets
Excellent toilets include baby changing and facilities for the disabled.

### Food
The area has kiosks and cafés.

### Seaside Activities
Everything from international events to Punch and Judy and beach games; swimming, sailing, windsurfing, water skiing; children's play areas.

### Wet Weather Alternatives
Sealife Centre, Brewers Quay and Timewalk, Nothe Fort, Deep Sea Adventure, Shire Horse Stables, Indoor Children's Play Centre, Weymouth Pavilion, Theatre, Model World, Tropical Jungle and Wessex Water Museum.

### Wildlife and Walks
The internationally recognised Lodmoor Nature Reserve and Radipole Swannery Reserve are less than five minutes' walk from the beach.

### Tourist Information
Weymouth Tourist Information Centre, The Esplanade, Weymouth, Dorset, DT4 8ED. Tel. 01305 785747

**Track Record**
**Weymouth PGGG**
**Lodmore GGGG**
Both beaches are EU designated.

South-West England

# DURDLE DOOR (EAST AND WEST)
## Dorset
*OS Refs: SY808803, SY804803*

Famous for Durdle Door Arch, one of the wonders of the British coastline created by the great erosive power of the sea (and probably the most photographed view along the Dorset coast), the beach is a narrow strand of mixed shingle, gravel and sand. The eastern end (Durdle Door Cove) is protected by the arch while the rest of the beach is partially sheltered by a submerged offshore reef. The beach is bounded at the western end by Bat's Head, a chalk headland. All the cliffs backing the beach are steep and prone to occasional rockfalls, so climbing or sheltering underneath them is not advised.

**Water Quality**
No routine sewage discharge has been identified.

**Bathing Safety**
As with most shingle and gravel beaches care must be taken since there can be a sudden steep slopes under the water. The western end of the beach can be cut off under certain tide and wave conditions.

**Litter**
Occasional marine litter, including plastic, rope and wood, is washed up in Durdle Door Cove.

**Access**
The beach is approached by a steep 800 metre long footpath from the cliff top car park. Access on to the eastern end of the beach is down a steep flight of steps cut into the bay cliff, and can be slippery in wet weather.

**Parking**
There is a large cliff-top car park at Durdle Door Caravan and Camping Park with excellent views across Weymouth Bay to the Isle of Portland.

**Toilets**
At the caravan park.

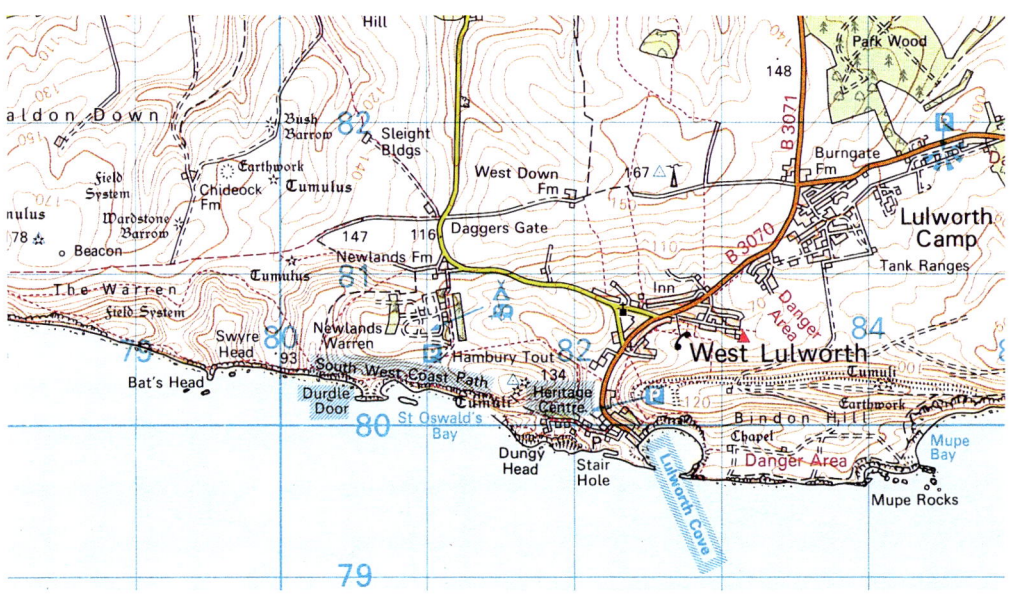

South-West England

*A testament to the power of wind and waves, the arch at Durdle Door shows the differing hardness of the rock strata along this stretch of coast.*

**Food**
Café and store in the caravan park.

**Seaside Activities**
Swimming, diving, snorkelling and fishing. The steep access to the beach means that heavy equipment should not be carried down.

**Wildlife and Walks**
The undulating cliffs form a challenging section of the Dorset Coast Path. To the east lies Lulworth Cove and the Isle of Purbeck, to the west White Nothe Headland. The reward for tackling this stretch of Heritage Coast is a fine view across Weymouth Bay and down to the glorious beaches below. The chalk habitat creates picturesque downland with its accompanying flora and fauna.

**Tourist Information**
Purbeck Tourist Information and Heritage Centre, South Street, Wareham, Dorset, BH20 4LU.
Tel. 01929 552740

**Track Record**
**Durdle Door East GGGG**
**Durdle Door West GGGG**
Both beaches are EU designated.

# KIMMERIDGE BAY
## Dorset
*OS Ref: SY907790*

This is an unusual beach of sand and fossil-bearing shale. Popular with divers, snorkellers, surfers and windsurfers, it is reached by a path from the cliff top car park.

 **Water Quality**
There is one private discharge.

 **Bathing Safety**
Normally safe; no lifeguard cover

 **Litter**
Litter bins are located in the car park.

Dogs are allowed on this beach.

 **Access**
From Corfe Castle follow Church Knowle road to Kimmeridge Village, then the toll road to the cliff-top car park. The path to the beach is not suitable for disabled visitors.

 **Parking**
Spaces for around 1,000 vehicles.

*A tower overlooks the peaceful bay at Kimmeridge.*

*South-West England*

 **Toilets**
Toilets are situated nearby.

 **Seaside Activities**
Swimming, surfing. Two slipways aid the movement of heavy equipment.

**Wet Weather Alternatives**
Swanage and Wareham offer plenty of activities for rainy days: contact the Tourist Information Centre.

 **Wildlife and Walks**
The South West Coastal Path: details from the Tourist Information Centre.

**Tourist Information**
See Durdle Door (previous page).

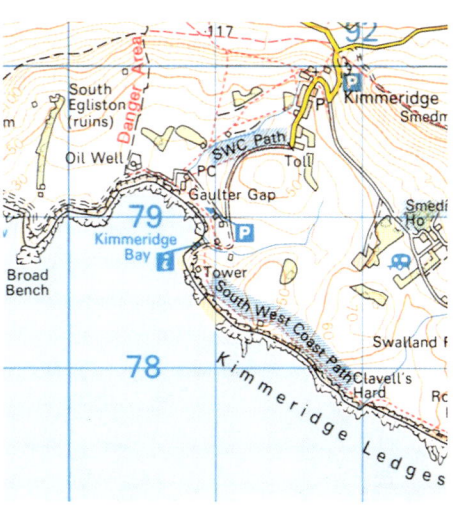

**Track Record FFPG**
EU designated beach.

59

# STUDLAND
## Dorset
*OS Ref: SZ035835*

Studland Bay has three miles of white sand backed by sand dunes. A naturist beach is at the east end of the bay. There are plenty of activities for rainy days in nearby Swanage, Poole and Bournemouth.

**Water Quality**
No routine sewage discharge has been identified.

**Bathing Safety**
Generally safe; trained lifeguards patrol the beach during the summer.

**Litter**
The beach is cleaned daily and dogs are allowed.

**Access**
Take the B3531 from Corfe Castle to Studland. Three car parks give access to the beach: the northernmost is flat and suitable for disabled visitors, while the others go through woods and via a steep hill respectively.

**Parking**
There are various car parks with a total capacity for some 2,000 cars.

South-West England

### Toilets
Toilets include baby changing and facilities for disabled visitors.

### Food
Cafés serve light hot and cold snacks.

### Seaside Activities
Swimming, sailing, windsurfing.

### Wet Weather Alternatives
There are a range of activities in nearby Swanage, Poole and Bournemouth. For further information contact the Tourist Information Centre.

### Wildlife and Walks
The South West Coast Path and the Studland Heath National Nature Reserve, behind the beach.

### Tourist Information
Swanage Tourist Information Centre, Shore Road, Swanage, Dorset.
Tel. 01929 422885

*Studland's fine white sand is blown and sculpted by the wind.*

**Track Record GGPG**
EU designated beach.

# SHELL BAY
## Dorset
*OS Ref: SZ038863*

Located at the very tip of the Studland Peninsula, at the mouth of Poole Harbour, Shell Bay is a beautiful beach. It is easily reached by road, or, for those who enjoy boat trips, by ferry from Sandbanks spit. The bay forms part of the Purbeck Heritage Coast and lies adjacent to the Studland Heath National Nature Reserve, providing plenty to occupy nature lovers. For those seeking the thrills of a traditional seaside resort, Swanage, Poole and Bournemouth are located conveniently close by.

### Water Quality
No routine sewage discharge has been identified.

### Bathing Safety
Bathing is considered to be generally safe, though strong currents can arise near the harbour entrance. The beach is patrolled between May and September from 8.00am to 8.30pm by wardens with first-aid training.

### Litter
The beach is cleaned daily. Dog fouling by-laws are in force.

### Access
By road from Studland or by ferry from Sandbanks.

### Parking
A car park is situated close to the beach.

### Toilets
These are located near the café.

### Food
There is a café at the entrance to the beach.

### Seaside Activities
Swimming, windsurfing, sailing and fishing. There is also a marine and watersports centre in the area.

### Wet Weather Alternatives
Poole and Sandbanks provide facilities including Tower Park Leisure Complex, four sports centres, Dolphin swimming pool, aquarium complex, Royal National Lifeboat Museum, Waterfront

and Scaplen's Court Museum, Arts Centre and Poole Pottery. Swanage is only a few miles away providing entertainment in the form of Swanage Steam Railway, Tithe Barn museum and the Vista swimming pool complex.

### Wildlife and Walks
The extensive Studland Heath National Nature Reserve backs Shell Bay. There is also a pedestrian ferry to Brownsea Island where tern colonies and lagoons can be seen. 80 hectares of this 202-hectare National Trust-owned island is a nature reserve run by the Dorset Trust for Nature

*Rolling dunes melt into light golden sand at Shell Bay.*

Conservation and possesses a varied range of habitats. Further information is available from the National Trust shop on the island's landing quay.

### Tourist Information
Swanage Tourist Information Centre, Shore Road, Swanage.
Tel 01929 422885

**Track Record GGGG**
EU designated beach.

# POOLE – SHORE ROAD
## Dorset
*OS Ref: SZ049885 (see map opposite)*

This narrow beach next to the main road serving domestic properties has a lighted promenade. There are scenic views across the harbour to Brownsea Island and the Purbeck Hills which provide an excellent backdrop for sunsets.

**Water Quality**
No routine sewage discharge has been identified.

**Bathing Safety**
Bathing is safe in these shallow waters. There is no lifeguard cover.

**Litter**
The beach is cleaned once a week during the summer season.

**Access**
The beach is well signposted from minor roads in the surrounding area. Disabled visitors will find level access to the promenade, but not on to the beach.

**Parking**
The promenade has space for 350 cars.

*Poole's pedestrian promenade is set against a steep pine- and shrub-covered backdrop.*

**Toilets**
Toilets include baby changing and facilities for disabled visitors.

**Food**
Nearby hotels and a public house serve light refreshments.

**Seaside Activities**
Swimming, jet-skiing, sailing, windsurfing, and harbour cruises.

**Wet Weather Alternatives**
See Sandbanks (opposite).

**Wildlife and Walks**
See Sandbanks (opposite).

**Tourist Information**
See Sandbanks (opposite).

**Track Record PPPG**
EU designated beach.

*South-West England*

# POOLE SANDBANKS
## Dorset
*OS Ref: SZ048880*

This popular and well managed beach has consistently good or excellent water quality. A fringe of soft golden sand stretches over five kilometres from the tip of Sandbanks Spit to merge with the beaches of Bournemouth. At the south-west end of the spit the beach is edged by dunes and overlooked by the Sandbanks Pavilion and recreation area.

**Water Quality**
No routine sewage discharge has been identified.

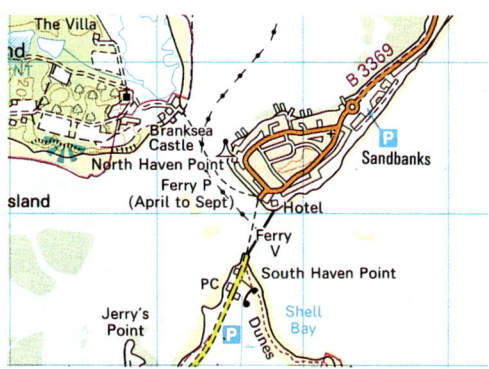

**Bathing Safety**
Safe, except at the extreme western end of the beach near the harbour entrance: warning signs indicate where not to swim. The beach is patrolled by lifeguards from May to September.

**Litter**
Cleaned daily; dogs are banned between May and September.

**Access**
The beach is reached from the adjacent B3369. Paths lead down the cliffs to the promenade.

**Parking**
Several car parks serve the beach.

**Toilets**
Toilets include mother and baby changing and facilities for the disabled.

**Food**
There are cafés and kiosks close by.

**Seaside Activities**
Swimming, windsurfing, sailing and fishing. Poole Harbour has several windsurfing and sailing schools. There is also a variety of children's amusements.

**Wet Weather Alternatives**
Facilities include Tower Park Leisure Complex, four sports centres, Dolphin swimming pool, aquarium complex, Royal National Lifeboat Museum, Waterfront and Scaplen's Court Museum, Arts Centre and Poole Pottery.

**Wildlife and Walks**
The three mile Evening Hill Discovery Trail runs adjacent to the beach. There is also a pedestrian ferry to Brownsea Island where tern colonies and lagoons can be seen. 80 hectares of this 202-hectare National Trust owned island is a nature reserve run by the Dorset Trust for Nature Conservation and possesses a varied range of habitats. Further information is available from the National Trust shop on the island's landing quay.

**Tourist Information**
Poole Tourism Services, The Quay, Poole, Dorset. Tel. 01202 253253

**Track Record GGGG**
EU designated beach.

# BOURNEMOUTH
## Dorset

Ranking as one of the most popular resorts on the south coast of England, Bournemouth can boast a range of attractions and facilities to cater for nearly every taste. The town also has some beautiful beaches, with golden sands often backed by majestic cliffs. Two in particular have been shown to have consistently good or excellent bathing water quality. These are:

Durley Chine, *OS Ref: SZ078903*
Fisherman's Walk (Southbourne), *OS Ref: SZ128913*

### Water Quality
No routine sewage discharges have been identified.

### Bathing Safety
Buoys mark the safe swimming area at Durley Chine. Bathing is generally safe at Fishermen's Walk. Both beaches are patrolled by voluntary or Council lifeguards during the summer.

### Litter
The beaches are cleaned daily during the summer.

### Access
For Durley Chine, take the first left at the end of the dual carriageway A338 westbound and follow the signposts. For Fisherman's Walk, from the A338 eastbound, proceed on to Castle Lane East (A3060) and follow the signs for Southbourne Beaches.

### Parking
Parking is available by the roadside in lower Durley Chine and along the over-cliff drive.

### Public Transport
Bournemouth Central is the nearest rail station; gas powered trains link the beaches to the town centre along the promenade. Yellow buses stop at the top of Durley Chine and at Fisherman's Walk.

### Toilets
Toilets, including facilities for dis-

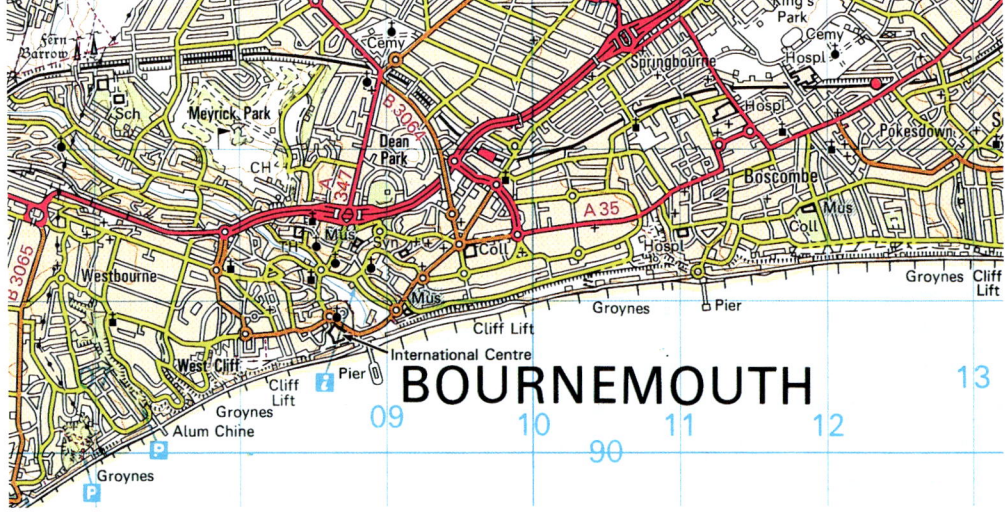

*South-West England*

abled visitors, are available at both beaches. There is also a mother and baby room situated at West Beach.

### Food

Kiosks at both beaches serve hot and cold snacks and a bar at the Durley Inn does hot and cold meals.

### Seaside Activities
Swimming and boating.

### Wet Weather Alternatives
Many facilities are located in the town centre within easy reach.

### Wildlife and Walks

Short walks can be taken along the wooded area adjacent to Durley Chine and

*The cliffs above Durley Chine afford views on to the sands of Fisherman's Walk.*

on the undercliff and overcliff paths at Fisherman's Walk. A three kilometre undercliff walk leads from Fisherman's Walk east to Hengistbury Head and there is a ten kilometre promenade leading west, linking Durley Chine with Sandbanks.

### Tourist Information
Bournemouth Visitor Information Bureau, Westover Road, Bournemouth, Dorset, BH1 2BU. Tel. 01202 451700

**Track Records**
**Durley Chine GGGG**
**Fisherman's Walk GGGG**
Both beaches are EU designated.

South-West England

# HENGISTBURY HEAD, EAST BOURNEMOUTH
## Dorset
*OS Ref: SZ170906*

This one-and-a-half kilometre long headland separates Poole Bay and Christchurch Bay and encloses Christchurch Harbour on its landward side. Largely undeveloped, it has been designated a Site of Special Scientific Interest because of the wide variety of plant and animal life it supports. There are two distinct beach areas: Hengistbury Head itself is a south-facing pebble beach below imposing sandstone cliffs. This three kilometre stretch of beach is undeveloped, in sharp contrast to the sand spit stretching north from the headland to the entrance of Christchurch Harbour where a string of beach huts face the groyne-ribbed sands. A ferry from Mudeford crosses the harbour entrance to drop passengers at the northern end of the beach where there is a quay and a short promenade.

### Water Quality
No routine sewage discharge has been identified.

### Bathing Safety
On the seaward-facing side of the headland bathing is safe except near the entrance to the harbour. Lifeguard cover is provided from Southbourne.

### Litter
Occasionally affected by marine litter from passing boats.

### Access
From the A338, follow the A3060 sign for Tuckton Bridge.

### Parking
There are 500 spaces at Hengistbury Head and a further 360 spaces at Mudeford Quay.

### Public Transport
Yellow buses (no 12 in summer and no 22 all year round) stop 5 minutes' walk from the beach. Paths lead from the grassed

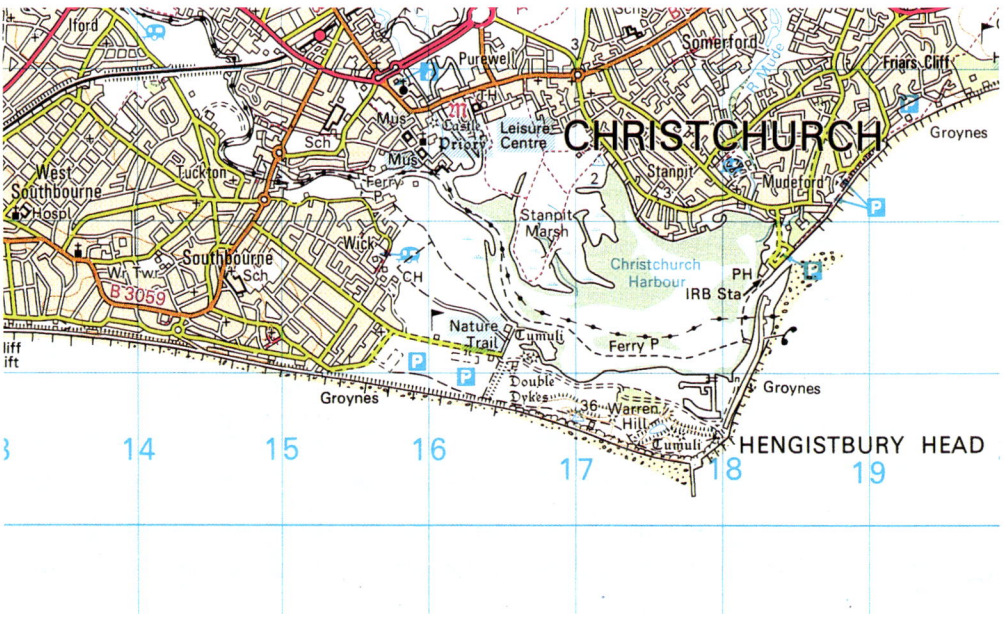

*South-West England*

*The south-facing pebble beach at Hengistbury is totally undeveloped – a picture of unspoilt beauty.*

areas at the end of Southbourne promenade to the beach.

 **Toilets**
Several blocks on the main beach include easy access facilities.

**Food**
Available from a beach café.

 **Seaside Activities**
Swimming and windsurfing; there is a windsurfing school nearby. An 18-hole golf course, land train and plenty of walks complete the attractions.

**Wildlife and Walks**
Hengistbury Head is of great archaeological interest with evidence of Iron Age and Roman settlement. Information signs along a nature trail interpret the history and wildlife of the area. An excellent viewpoint at OS Ref. SZ178904 gives splendid views.

**Tourist Information**
Bournemouth Visitor Information Bureau, Westover Road, Bournemouth, Dorset, BH1 2BU. Tel. 01202 451700

**Track Record GGPG**
EU designated beach.

# MUDEFORD SANDBANK, CHRISTCHURCH
## Dorset
*OS Ref: SZ183912*

This sand spit bounded by Christchurch Harbour on one side and Christchurch Bay on the other has beaches on both flanks: the one on the harbour side is mainly of compressed shingle and not recommended for swimming. The beach on the seaward side, by contrast, is sandy and the water is ideal for bathing.

**Water Quality**
No routine sewage discharge has been identified.

**Bathing Safety**
Normally safe; there is no lifeguard cover.

**Litter**
The beach is cleaned daily. A poop-scoop scheme was introduced in spring 1997.

**Access**
There is no vehicular access to the beach but a land train service runs from Hengistbury Head where there are parking facilities. Ferries run across the harbour from both Mudeford Quay and Christchurch Town Quay.

**Parking**
See Hengistbury Head (previous page).

**Toilets**
There are several including baby changing and facilities for disabled visitors.

**Food**
A beach café sells a wide variety of hot and cold food and beverages.

**Seaside Activities**
Swimming, boating, windsurfing.

**Wet Weather Alternatives**
Bournemouth and Christchurch offer a wide range of wet weather activities: contact the Tourist Information Centre for further information.

**Wildlife and Walks**
Hengistbury Head and Stawpit Marsh are important wildlife conservation areas; a warden's headquarters has information on trails through the reserves. Further guidebooks and leaflets are available from the Tourist Information Centre.

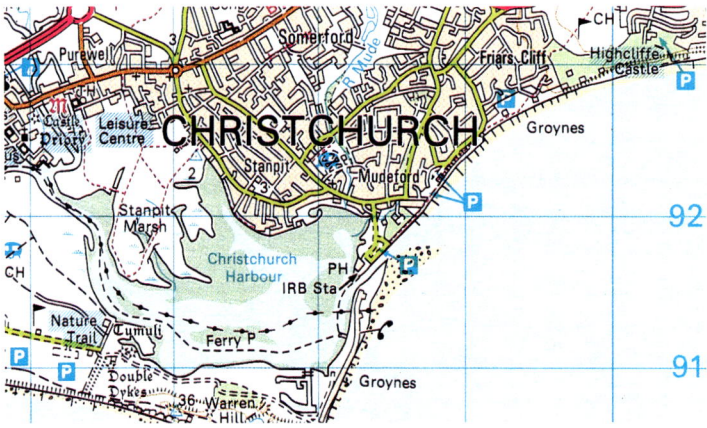

**Tourist Information**
Christchurch Tourist Information Centre,
23 High Street,
Christchurch,
Dorset, BH23 1AB.
Tel. 01202 471780

**Track Record PPPG**
EU designated beach.

*South-West England*

# ST HELENS
## Isle of Wight
*OS Ref: SZ637892*

This quiet rural beach of sand and shingle is backed by land managed by the National Trust, providing excellent walking and bird-watching opportunities.

**Water Quality**
No routine sewage discharge has been identified.

**Bathing Safety**
Generally safe. There is no lifeguard cover but basic lifesaving equipment is situated on the Esplanade.

**Litter**
The beach is cleaned daily; litter bins are located on the Esplanade.

**Access**
The beach is easily reached from Dover Road in St Helens Village. Access to the beach from the promenade is via a public slipway and timber steps.

**Parking**
There are spaces on the promenade and on National Trust property.

**Toilets**
These include facilities for disabled visitors.

**Food**
There is a restaurant on the Esplanade.

**Seaside Activities**
Swimming and jet skiing.

**Wet Weather Alternatives**
In Ryde there is a ten-pin bowling alley, ice rink, indoor swimming pool, leisure centre, golf driving range and flamingo park.

**Wildlife and Walks**
The Isle of Wight Coastal Path. For further details of walks and the wildlife of the Isle of Wight contact the Tourist Information Centre.

**Tourist Information**
Ryde Tourist Information Centre, 81 Union Street, Ryde, Isle of Wight. Tel. 01983 562905

**Track Record PPPG**
EU designated beach.

## South-West England

| RATING | NAME | TRACK RECORD | SEWAGE OUTLET | REMARKS |
|---|---|---|---|---|
| | SOMERSET | | | |
| | **Clevedon** | | | |
| ~ | Bay | ~~~~ | ■ | ▲ ■ |
| P | Swimming pool - EU ST398712 | FPFP | ■ | ▲ ■ |
| | **Weston-super-Mare** | | Screened and disinfected, 75,000, remote. | ↑ 2000 |
| P | Uphill Slipway - EU ST312588 | FFPP | | ■ |
| ~ | Sanatorium ST314600 | FFP~ | | ■ |
| P | Main Beach - EU ST316607 | FPPP | | ■ 🗑 Rocky coves at north end. |
| ~ | Grand Pier ST317615 | FFP~ | | ■ |
| ~ | Near Marine Lake ST312619 | PF~~ | | ■ Rocks at north end. |
| P | Kewstoke Sand Bay (near Weston) - EU ST330635 | FPFP | | ■ |
| | **Berrow** | | | |
| P | North - EU ST293545 | PFPP | ■ | ■ |
| ~ | South ST290535 | F~~~ | ■ | ■ |
| P | **Brean** - EU ST296585 | PPPP | ■ | ■ |
| | **Burnham-on-Sea** | | | |
| ~ | Yacht Club ST301480 | PF~~ | ■ | ■ |
| P | Jetty - EU ST302488 | PPPP | Screened and disinfected, 36,000, at HWM | ■ 🗑 Bathing mostly dangerous. |
| ~ | **East Quantoxhead** | ~~~~ | Screened, 800, at LWM. | ▲ ■ |
| F | **Doniford** | ~~~F | Screened, 5,000, 100m above LWM. | ■ ■ ↑ 1997 |
| P | **Watchet** | ~~~P | Raw, 4,500, 100m above LWM. | ■ ■ ↑ 1999 |
| P | **Blue Anchor** - EU ST023435 | PPPP | ■ | ■ ■ |
| | **Dunster** | | | |
| P | North West - EU SS997455 | PPPP | ■ | ■ ■ |
| ~ | South East | ~~~~ | ■ | ■ ■ |
| | **Minehead** | | | |
| P | Terminus - EU SS973465 | PPPP | Screened, 800m below LWM | ■ ↑ 1999 |
| ~ | The Strand SS978463 | P~~~ | ■ | ■ |
| | **Porlock Bay** | | | |
| P | Porlock Weir - EU SS864479 | GGGP | 3, all raw, 2,500 (total), all at LWM. | ■ ↑ 1997 |

72  ■ Sand    ■ Shingle    ■ Pebbles    ▲ Rocks    ■ Mud    ? No information supplied

*South-West England*

| RATING | NAME | TRACK RECORD | SEWAGE OUTLET | REMARKS |
|---|---|---|---|---|
| | DEVON (NORTH) | | | |
| F | Lynmouth - EU SS725497 | FFFF | Secondary, UV, 4,300, 110m below LWM. | 1997 |
| P | Wringcliff SS700497 | ~~~P | | |
| P | Woody Bay SS678490 | ~~~P | | |
| F | Combe Martin - EU SS577473 | PFPF | Raw, 3,600, 65m below LWM. | |
| | Ilfracombe | | Secondary, 22,744, 235m and 30m below LWM. | |
| F | Hele Beach - EU SS535479 | PFFF | | |
| F | Capstone Beach - EU SS518479 | FPFF | | |
| P | Tunnels - EU SS514478 | PPPP | | |
| | Mortehoe, Rockham Bay SS458460 | ~~G~ | | |
| ~ | Barricane Bay | ~P~~ | | Surrounded by rocks. |
| G | Woolacombe Village Beach - EU SS456437 | GGPG | Secondary, 13,200, 100m below LWM. | FEATURED |
| G | Putsborough Beach - EU SS447408 | GGFG | | FEATURED |
| P | Croyde Bay - EU SS434393 | PPGP | Screened, 6,400, below LWM. | 1999 Strong undertow at all times. |
| G | Saunton Sands - EU SS445376 | PPPG | | |
| P | Instow - EU SS471304 | FFPP | Screened, 8,500, at LWM. | 1999. Polluted by rivers Taw andTorridge. |
| P | Westward Ho! - EU SS432294 | PPPP | Screened, 4,600, 10m below LWM. | 1999 |
| ~ | Clovelly SS318249 | ~~G~ | Raw, 1,300, at LWM. | 2005 Sand at low tide only. |
| ~ | Shipload Bay SS2428 | ~G~~ | | Bathers beware of currents. Access difficult. |
| G | Hartland Quay - EU SS223248 | GGGG | | FEATURED |
| ~ | Welcombe Mouth SS213181 | ~G~~ | | Some sand at low tide. |
| | CORNWALL | | | |
| | Bude | | | |
| G | Sandy Mouth - EU SS202099 | PGGG | | FEATURED |
| G | Crooklets - EU SS203072 | PPGG | | FEATURED |
| P | Summerleaze - EU SS204066 | FPPP | Primary, 12,700, LSO. | Backed by pebble ridge. |
| G | Widemouth Bay - EU SS198024 | GPGG | | FEATURED |

■ No discharge identified  ↑ Improvements planned  ⁇ Insufficient information to feature  ▓ Cleaned regularly

## South-West England

| RATING | NAME | TRACK RECORD | SEWAGE OUTLET | REMARKS |
|---|---|---|---|---|
| ~ | Crackington Haven SX142969 | F~~~ | ■ | □ |
| ~ | Boscastle SX0991 | ~~~~ | Raw, 1,300, at LWM. | ↑ 2005 Deep narrow fjord-like harbour. |
| ~ | Tintagel | ~~~~ | Raw, 1,500, at LWM. | ↑ 2005 ■ |
| ~ | Trebarwith Strand SX048863 | F~~~ | ■ | □ ▲ ?? Swimming can be dangerous. |
| G | Tregardock SX040840 | ~~~G | ■ | □ ▲ Bathing unsafe for beginners. |
| ~ | Port Isaac | ~~~~ | Secondary, 1,800, at LWM. | Fishing port. |
| G | Polzeath - EU SW936792 | PPGG | Screened, 1,700, at LWM. Flow transferred 1997. | □ ♨ ■ |
| G | Greenaway Beach SW929786 | ~~~G | (see Polzeath) | ?? |
| G | Daymer Bay - EU SW928776 | PGGG | ■ | FEATURED |
| P | Rock Beach - EU SW927758 | PGPP | Secondary, 1,100, at LWM Flow transferred 1997. | □ |
| P | Padstow SW919764 | P~~P | Screened, 3,800. Flow transferred 1997. | Small harbour. No swimming. |
| P | Trevone Bay - EU SW892761 | FPPP | Raw, 1,000, at LWM. Flow transferred 1997. | □ Bathing can be dangerous. |
| G | Harlyn Bay - EU SW877755 | PGPG | Secondary, from Cataclews. | □ Not featured due to adjacent discharge. |
| G | Mother Ivey's Bay - EU SW863760 | GGGG | ■ | ?? □ Not featured due to adjacent discharge. |
| G | Porthbeor SW862320 | ~~~G | ■ | ?? |
| G | Constantine Bay - EU SW857746 | GGGG | ■ | FEATURED |
| G | Treyarnon Bay - EU SW857740 | GGGG | ■ | FEATURED |
| P | Porthcothan - EU SW857720 | ~~PP | ■ | □ With sand dunes. |
| F | Mawgan Porth - EU SW848674 | FPPF | Secondary, UV, from St Columb Major. | □ Surfing dangerous at low tide. |
| ~ | Bedruthan Steps | ~~~~ | ■ | □ ▲ Bathing dangerous. |
| P | Watergate Bay - EU SW841649 | PPGP | ■ | □ |
| P | Lusty Glaze - EU SW823624 | ~~~P | ■ | □ |
| P | Porth Beach - EU SW829627 | P~~P | ■ | □ |
|  | Newquay |  | Screened, 50,000, 75m below LWM. Discharge from Towan Head. | ↑ 1999 |
| P | Great Western - EU SW815618 | ~~~P |  | □ |
| G | Tolcarne Beach - EU SW810620 | ~~~G |  | □ Not featured due to adjacent discharge. |
| G | Towan Beach - EU SW810620 | PPPG |  | □ Not featured due to adjacent discharge. |

74   □ Sand      Shingle      Pebbles     ▲ Rocks     ■ Mud     ? No information supplied

South-West England

| RATING | NAME | TRACK RECORD | SEWAGE OUTLET | REMARKS |
|---|---|---|---|---|
| G | Fistral Beach - EU SW796623 | GGGG | | 🟨 🗑 Strong currents when rough. Not featured due to adjacent discharge. |
| P | **Crantock** - EU SW784608 | PPGP | 🟩 | 🟨 Swimming dangerous at low tide and near estuary. |
| P | **Holywell Bay** - EU SW765595 | PPPP | 🟩 | 🟨 Steep shelf makes surfing dangerous at high tide. |
| G | **Perranporth** Penhale Sands - EU SW762570 | GGGG | 🟩 | FEATURED |
| G | Village End Beach - EU SW757548 | PPPG | Secondary, disinfected, 12,000, at LWM. | FEATURED |
| P | **Trevaunance Cove** - EU SW723517 | FFPP | Screened, 4,000, 2 km to the west, at LWM. | 🟨 ⬛ |
| G | **Porthtowan Sandy** - EU SW691481 | GGGG | 🟩 | 🗑 Not featured due to sewage related discharge. |
| G | **Portreath** - EU SW653455 | GGGG | Screened, 22,000, at LWM east of village. | 🔺 2000 🟨 Not featured due to adjacent discharge. |
| ~ | **North Cliffs** (Deadman's Cove, Cambourne) SW625434 | ~~~~ | Raw, 23,000, at LWM. | 🔺 🟨 ⛰ |
| P | **The Towans** Godrevy - EU SW581417 | PPGP | 🟩 | 🟨 Swimming dangerous at low tide. |
| G | Hayle, St Ives - EU SW563395 | GGGG | 🟩 | FEATURED |
| G | **Carbis Bay** Porth Kidney Sands - EU SW540385 | GPGG | 🟩 | FEATURED |
| P | Station Beach - EU SW528389 | GPGP | 🟩 | 🟨 |
| G | **St Ives** Porthminster - EU SW522402 | PFPG | 🟩 | FEATURED |
| G | Porthgwidden - EU SW522411 | FFFG | 🟩 | FEATURED |
| G | Porthmeor - EU SW515410 | GPGG | 🟩 | FEATURED |
| G | **St Just Priest's Cove** SW352317 | P~~G | 🟩 | FEATURED |
| P | **Whitesand Bay** Sennen Cove - EU SW355264 | GGGP | Raw, 700. | 🔺 2005 🟨 ⛰ |
| ~ | **Porthgwarra** | ~~~~ | Primary, 12, septic tank. | 🟨 |
| G | **Porthcurno** - EU SW387223 | PGGG | Macerated, 200, at LWM. | 🔺 2005 🟨 🗑 Not featured due to adjacent discharge. |
| ~ | **Lamorna Cove** SW4524 | ~~~~ | Tidal tank, at LWM. | 🟨 ⛰ |

🟩 No discharge identified   🔺 Improvements planned   ⁇ Insufficient information to feature   🗑 Cleaned regularly

## South-West England

| RATING | NAME | TRACK RECORD | SEWAGE OUTLET | REMARKS |
|---|---|---|---|---|
| F | **Mousehole** SW470263 | F~~F | ■ | Fishing port. |
| P | **Marazion and Mounts Bay** Wherrytown - EU SW467294 | FFPP | ■ | |
| P | Heliport - EU SW485311 | FFPP | ■ | |
| G | Penzance - EU SW475298 | FFPG | ■ | ?? |
| P | Little Hogus - EU SW513310 | FPPP | ■ | |
| G | **Perran Sands** - EU SW539293 | GGGG | Raw, 1,000, at LWM 0.5km west of beach. | Not featured due to adjacent discharge. |
| P | **Praa Sands** West - EU SW577281 | GGPP | ■ | |
| G | East - EU SW585276 | GGGG | ■ | FEATURED |
| P | **Rinsey Head** SW593269 | ~~~P | ■ | Granite headland. |
| F | **Porthleven West** - EU SW632253 | FFFF | Raw, 3,500, 50m below LWM. | ↑ 1997 Flint. Bathing dangerous. |
| G | **Jangye Ryne** SW659207 | ~~~G | ■ | ?? |
| G | **Gunwalloe Cove** - EU SW654225 | GGGG | ■ | Not featured due to adverse reports. |
| P | **Poldhu Cove** - EU SW665198 | PPGP | ■ | Bathing dangerous at low tide. |
| P | **Polurrian Cove** - EU SW668187 | PPPP | Macerated, 1,600, at LWM. | ↑ 2005 |
| ~ | **Kynance Cove** SW683132 | ~~G~ | ■ | |
| P | **Polpeor** | G~~P | ■ | Cove. |
| F | **Lizard Church Cove** SW661705 | P~~F | Raw, 1,600, 500m below LWM. | ↑ 2005 Fishing cove. |
| P | **Kennack Sands, Kuggar** - EU SW734165 | GGGP | ■ | |
| P | **Coverack** - EU SW783186 | PFPP | Primary, 800, 100m below LWM. | |
| G | **Porthoustock** - EU SW807217 | GGGG | ■ | Not featured due to adverse reports. |
| F | **Porthallow** - EU SW797233 | FFFF | Raw, 100. | Grey stones. |
| P | **Maenporth** - EU SW790296 | FPPP | ■ | |
| G | **Falmouth** Swanpool Beach - EU SW790233 | FGGG | ■ | Not featured due to adverse reports. |
| G | Gyllyngvase - EU SW809316 | PGGG | Macerated, 9000, off Pendennis Point. | ↑ 1997 Not featured due to adjacent discharge. |
| ~ | Feock Loe Beach SW826320 | P~~~ | (See Gyllyngvase) | ↑ 1997 |

 Sand  Shingle  Pebbles  Rocks  Mud  ? No information supplied

# South-West England

| RATING | NAME | TRACK RECORD | SEWAGE OUTLET | REMARKS |
|---|---|---|---|---|
| ~ | **St Mawes** SW849331 | ~G~~ | Secondary, 1000, deep water below LWM. | 🟨 |
| ~ | **St Antony's Head** | ~~~~ | 🟩 | Fine shingle at low tide. |
| ~ | **Towan Beach** SW870329 | ~~G~ | 🟩 | 🟨 ⛰ |
| ~ | **Portscatho** | ~~~~ | 2, raw, 142 (total), both at LWM. | ⬆ 2005 🟨 ⛰ |
| ~ | **Porthcurnick Beach** | ~G~~ | 🟩 | 🟨 ⛰ |
| G | **Pendower Beach** - EU SW898381 | PPGG | 🟩 | ❓ 🟨 ⛰ |
| ~ | **Carne Beach** (Pendower) | G~~~ | 🟩 | ⛰ |
| ~ | **Portloe** SW938394 | FF~~ | 2, macerated, 200 (total), both at LWM. | ⬆ 2005 🟨 ⛰ Fishing village. |
| ~ | **Portholland Beach** | ~F~~ | 🟩 | 🟨 Sand at low tide. |
| P | **Porthluney Cove** - EU SW973413 | PPPP | 🟩 | 🟨 |
| ~ | **Hemmick Beach** SW9940 | ~G~~ | 🟩 | 🟨 Small, sandy bay. |
| G | **Bow or Vault Beach, Gorran Haven** - EU SX010408 | GGGG | Discharge from Gorran Haven Beach nearby. | ❓ 🟨 |
| F | **Little Perhaver** - EU SX013417 | PPPF | Secondary and disinfected, 2600, at LWM south of village. | 🟨 |
| P | **Portmellon** - EU SX016439 | PPPP | 🟩 | 🟨 |
| ~ | **Mevagissey** SX016439 | FF~~ | Screened, 3,000, at LWM east of harbour. | ⬆ Fishing harbour. No beach. |
| P | **Polstreath** - EU SX017454 | PPPP | 🟩 | 🟨 Steep cliffs. |
| P | **Pentewan** - EU SX018467 | GPPP | Discharge from Menagwins. | 🟨 |
| G | **Porthpean** - EU SX032507 | GGGG | 🟩 | FEATURED |
| G | **Duporth Beach** - EU SX035513 | ~~~G | 🟩 | ❓ |
| G | **Charlestown** - EU SX042516 | GGGG | 🟩 | 🟨 Not featured at request of owners. |
| G | **Crinnis Beach** Golf links - EU SX063522 | PGGG | 🟩 | ❓ 🟨 |
| G | Leisure Centre - EU SX056521 | GGGG | 🟩 | ❓ 🟨 Swimming dangerous near stream. |
| P | **Par Sands** - EU SX083533 | PGGP | Secondary, 21,000, LSO. | 🟨 |
| P | **Polkerris** - EU SX092521 | PGPP | Screened, 60, 5m below LWM. | 🟨 |
| ~ | **Polridmouth Beach** SX1050 | ~P~~ | 🟩 | 🟨 Sheltered from south-westerly winds. |

🟩 No discharge identified   ⬆ Improvements planned   ❓ Insufficient information to feature   🗑 Cleaned regularly

| South-West England | | | | |
|---|---|---|---|---|
| RATING | NAME | TRACK RECORD | SEWAGE OUTLET | REMARKS |
| P | **Fowey** Readymoney Cove - EU SX118511 | FPPP | Secondary, 2,500, below LWM. Also stormwater. | |
| ~ | **Lantic Bay** | ~P~~ | | Strong undertow. |
| ~ | **Lansallos Bay** | ~~~~ | | |
| ~ | **Polperro** SX210509 | FF~~ | Macerated, 3,500, at LWM. | |
| P | **Talland Bay** SX223515 | ~~~P | | Cliffs |
| P | **Looe - Hannafore** SX256526 | ~~~P | | |
| F | **Looe - Plaidy** SX263537 | ~~~F | | |
| F | **East Looe** - EU SX257532 | FPPF | | |
| P | **Millendreath** - EU SX257532 | PPPP | | |
| P | **Seaton Beach** - EU SX303543 | PGPP | Raw, 1,300, at LWM. | 1997 |
| G | **Downderry** - EU SX314538 | PGGG | Raw, 1,000, below LWM east of beach. | 1997 Not featured due to adjacent discharge. |
| G | **Portwrinkle, Freathy** Whitsand Bay - EU SX359538 | PPGG | Secondary, 800, below LWM. | Backed by cliffs. Bathing is unsafe. |
| ~ | **Cawsand Bay** SX434502 | ~~F~ | 3, raw, 260, at LWM. | 2005 |
| ~ | **Kingsands Bay** SX4350 | ~~~~ | 2, raw, 440, 5m and 12m below LWM. | 2005 |
| | DEVON (SOUTH) | | | |
| F | **Plymouth Hoe** West - EU SX475537 | FFPF | (See below) | |
| F | East - EU SX478537 | FFPF | 23, primary and raw, 100,000 below LWM. | |
| P | **Bovisand Bay** - EU SX493505 | FGPP | | |
| P | **Wembury** - EU SX516485 | FPPP | Secondary, 4,400, 100m below LWM | |
| G | **Row Cove Beach** SX566466 | ~~~G | | |
| P | **Mothecombe** - EU SX610473 | PFGP | | Bathing safe only on incoming tide. |
| G | **Challaborough** - EU SX649448 | GGGG | | FEATURED |
| P | **Bigbury-on-Sea** North - EU SX649443 | GGGP | | |
| P | South - EU SX615441 | PPPP | Secondary, 1,260, 50m below LWM. | Swimming dangerous near river mouth. |

Sand    Shingle    Pebbles    Rocks    Mud    No information supplied

South-West England

| RATING | NAME | TRACK RECORD | SEWAGE OUTLET | REMARKS |
|---|---|---|---|---|
| P | **Bantham** - EU<br>SX662438 | PGGP | 🟩 | 🟨 🗑 Beach backed by dunes. |
| P | **Mouthwell Sands**<br>SX675401 | ~~~P | 🟩 | 🟨 |
| G | **Thurlestone**<br>   North - EU<br>SX674421 | GGGG | 🟩 | FEATURED |
| G |    South Milton Sands - EU<br>SX676417 | GGGG | 🟩 | FEATURED |
| F | **Wonwell Sands**<br>SX617473 | ~~~F | 🟩 | 🟨 |
| P | **Hope Cove** - EU<br>SX675397 | PPPP | 🟩 | 🟨 ⛰ 🗑 Beach cleaned during bathing season. |
| ~ | **Soar Mill Cove** | ~P~~ | 🟩 | 🟨 ⛰ Cliffs. |
| F | **Salcombe South Sands** - EU<br>SX728377 | FPPF | Secondary, 3500, below LWM. | 🟨 |
| G | **Salcombe North Sands** - EU<br>SX731382 | PPPG | 🟩 | FEATURED |
| G | **Mill Bay** (Nr Salcombe) - EU<br>SX740382 | GGGG | 🟩 | FEATURED |
| G | **Abrahams Hole**<br>SX754368 | ~~~G | 🟩 | ⁉️ |
| G | **Gammon Head**<br>SX766358 | ~~~G | 🟩 | ⁉️ |
| G | **Great Mattiscombe**<br>SX816396 | ~~~G | 🟩 | ⁉️ |
| ~ | **Hallsands** | ~G~~ | 🟩 | 🟧 |
| G | **Beesands**<br>SX820405 | ~P~G | 🟩 | FEATURED |
| G | **Torcross** - EU<br>SX823419 | GGGG | Raw, 300, at LWM south of village. | Fishing port. Not featured due to adjacent discharge. |
| G | **Slapton Sands** - EU<br>SX828440 | GGPG | 🟩 | FEATURED |
| G | **Strete Gate Beach**<br>SX835455 | ~~~G | 🟩 | ⁉️ |
| G | **Blackpool Sands**<br>   Stoke Fleming - EU<br>SX855478 | GGGG | 🟩 | FEATURED |
| ~ | **Leonard's Cove** | ~~~~ | Raw, 800, at LWM. | ⬆ 2005 🟧 |
| P | **Dartmouth Castle and Sugary Cove** - EU<br>SX886502 | PPGP | 5, macerated, 11,900 at LWM to 50m below. | ⬆ 1997 🟧 |
| P | **St Mary's Bay** - EU<br>SX932551 | PPPP | Screened, 90,000, 220m below LWM. | ⬆ 2005 🟨 🪨 |
| G | **Shoalstone Beach** - EU<br>SX932566 | GGPG | 🟩 | FEATURED |
| ~ | **Churston Cove**<br>SX919569 | PPG~ | 🟩 | ⁉️ 🟧 |
| P | **Broadsands Beach** - EU<br>SX897574 | GFPP | 🟩 | 🟨 🪨 🟫 |
| P | **Goodrington Sands** - EU<br>SX893594 | PPPP | Stormwater. | ⬆ 1997 🟨 🪨 |

🟩 No discharge identified    ⬆ Improvements planned    ⁉️ Insufficient information to feature    🗑 Cleaned regularly

## South-West England

| RATING | NAME | TRACK RECORD | SEWAGE OUTLET | REMARKS |
|---|---|---|---|---|
| | **Paignton** | | | |
| P | Paignton Sands - EU SX894606 | PFPP | Stormwater. | ⬆ 1997 |
| P | Preston Sands - EU SX896617 | PFPP | Stormwater | ⬆ 1997 |
| P | **Hollicombe** - EU SX898621 | PPPP | 🟩 | |
| G | **Elbury Cove** SX903570 | ~~~G | 🟩 | ❓ |
| G | **Torre Abbey Sands** - EU SX909635 | PPPG | 🟩 | FEATURED |
| G | **Beacon Cove** - EU SX919630 | PGGG | 🟩 | FEATURED |
| G | **Meadfoot Beach** - EU SX930630 | GPGG | 🟩 | FEATURED |
| G | **Anstey's Cove (Redgate Beach)** - EU SX930648 | PPGG | 🟩 | FEATURED |
| G | **Babbacombe Beach** - EU SX930654 | PPGG | 🟩 | FEATURED |
| G | **Oddicombe Beach** - EU SX926657 | GGGG | 🟩 | FEATURED |
| G | **Watcombe Beach** - EU SX926673 | PGGG | 🟩 | FEATURED |
| G | **Maidencombe Beach** - EU SX927685 | ~GGG | 🟩 | FEATURED |
| G | **Ness Cove** - EU SX938717 | ~GGG | 🟩 | FEATURED |
| P | **Shaldon** - EU SX935723 | PPPP | Secondary, 50,000, LSO. | |
| P | **Teignmouth** - EU SX943728 | PPPP | | |
| G | **Holcombe** - EU SX956746 | PFPG | 🟩 | ❓ |
| | **Dawlish** | | | |
| P | Town - EU SX965768 | PPPP | Raw, 12,000, 100m below LWM. | ⬆ 2000 Cliffs |
| P | Coryton Cove - EU SX961760 | PPPP | (See Dawlish Town) | |
| G | Dawlish Warren - EU SX983787 | PGGG | (See Dawlish Town) | Not featured due to adjacent discharge. |
| G | **Exmouth** - EU SY009799 | FPGG | Secondary, 42,700, 170m below LWM. | FEATURED |
| P | **Sandy Bay** - EU SY033798 | FPPP | 🟩 | |
| P | **Budleigh Salterton** - EU SY069819 | FFPP | 🟩 | |
| P | **Ladram Bay** - EU SY119895 | PPPP | 🟩 | |
| P | **Sidmouth** - EU SY127872 | PFPP | Macerated and screened, 14,000, 400m below LWM. | ⬆ 2000 Not featured due to adjacent discharge. |
| G | Jacob's Ladder - EU SY119869 | PPGG | (See Sidmouth) | Not featured due to adjacent discharge. |

 Sand   Shingle   Pebbles   Rocks   Mud  ❓ No information supplied

*South-West England*

| RATING | NAME | TRACK RECORD | SEWAGE OUTLET | REMARKS |
|---|---|---|---|---|
| ~ | **Branscombe** | G~~~ | ■ | 🗑 |
| G | **Beer** - EU SY231891 | PFPG | Raw, 2,000, at LWM 1km south of the town. | ⬆ 2005 🗑 Not featured due to adjacent discharge. |
| P | **Seaton** - EU SY245898 | GPPP | Secondary, 11,000, into estuary. | 🗑 Steep beach. |
| | DORSET | | | |
| ~ | **Lyme Regis** Monmouth Beach SY337915 | F~~~ | ■ | 🗑 Slight shelving. |
| P | Church Beach - EU SY343921 | P~PP | ■ | ■ 🗑 |
| G | Cobb/Town Beach - EU SY339918 | FPPG | Disinfected secondary, UV, 6,000, 400m below LWM. | FEATURED |
| G | **Charmouth** West - EU SY363930 | PPPG | (See Charmouth East) | ■ 🗑 Not featured due to adjacent discharge. |
| ~ | East SY367929 | FPF~ | Screened, 8000, LSO | ■ 🗑 Slight shelving. |
| G | **Seatown** - EU SY418916 | PPGG | Secondary 2000. | FEATURED |
| G | **Eypemouth** - EU SY446910 | GPGG | ■ | FEATURED |
| G | **West Bay** (West) - EU SY459904 | PGGG | Screened, 30,000, LSO. | ⬆ 2000 ■ 🗑 Not featured due to adjacent discharge. |
| ~ | **Burton Bradstock** SY490887 | PGG~ | ■ | ■ Steeply shelving. |
| ~ | **Chesil Cove** SY682735 | PGG~ | Screened, 86,000, LSO. | ⬆2000 ■ |
| G | **Portland Harbour** Sandsfoot - EU SY673772 | FPGG | ■ | ■ ■ [??] |
| G | Castle Cove - EU SY676775 | PPGG | ■ | ■ Not featured at the request of the owner. |
| P | **Weymouth** South SY682789 | PP~P | ■ | ■ |
| G | Central - EU SY681794 | PGGG | ■ | FEATURED |
| ~ | **Lodmoor West** SY687806 | GG~~ | ■ | ■ 🗑 |
| G | **Lodmoor** - EU SY688807 | GGGG | ■ | FEATURED |
| G | **Overcome** SY692812 | ~~~G | ■ | [??] |
| G | **Church Ope Cove** - EU SY697710 | GGGG | ■ | ■ 🗑 ▲ ■ [??] |
| F | **Bowleaze** - EU SY704818 | PPPF | ■ | ■ 🗑 |
| P | **Ringstead Bay** - EU SY751813 | FPGP | ■ | ■ 🗑 |

■ No discharge identified   ⬆ Improvements planned   [??] Insufficient information to feature   ▥ Cleaned regularly

## South-West England

| RATING | NAME | TRACK RECORD | SEWAGE OUTLET | REMARKS |
|---|---|---|---|---|
| G | **Durdle Door**<br>West - EU<br>SY804803 | GGGG | ■ | FEATURED |
| G | East - EU<br>SY808803 | GGGG | ■ | FEATURED |
| G | **Lulworth Cove** - EU<br>SY824799 | PPPG | Screened, 2,000, below LWM outside cove. | Not featured due to adjacent discharge. |
| ~ | **Worbarrow Bay** | ~~~~ | ■ | |
| G | **Kimmeridge Bay** - EU<br>SY907790 | FFPG | Private discharge. | FEATURED |
| ~ | **Swanage South**<br>SZ031788 | FFF~ | Macerated, 12,000, 100m below LWM. | ⬆ 1999 |
| G | **Swanage Central** - EU<br>SZ032791 | FPPG | (See Swanage South) | Not featured due to adjacent discharge. |
| ~ | **Swanage North**<br>SZ031797 | FFF~ | (See Swanage South) | Beach backed by promenade and road. |
| G | **Studland** - EU<br>SZ035835 | GGPG | ■ | FEATURED |
| G | **Shell Bay (Poole)** - EU<br>SZ038863 | GGGG | ■ | FEATURED |
| | **Poole** | | | |
| P | Rockley Sands - EU<br>SY972908 | PPPP | ■ | |
| P | Lake - EU<br>SY983904 | PPPP | ■ | |
| G | Shore Road - EU<br>SZ049885 | PPPG | ■ | FEATURED |
| ~ | Branksome Chine<br>SZ066897 | PPG~ | ■ | Adjacent to cliffs. |
| G | Sandbanks - EU<br>SZ048880 | GGGG | ■ | FEATURED |
| | **Bournemouth** | | | |
| G | Durley Chine - EU<br>SZ078903 | GGGG | ■ | FEATURED |
| ~ | Alum Chine<br>SZ076903 | GG~~ | ■ | |
| G | Bournemouth Pier - EU<br>SZ088906 | PFPG | Stormwater. | Not featured due to adjacent discharge. |
| P | Boscombe Pier - EU<br>SZ112911 | FPPP | Stormwater. | |
| G | Fisherman's Walk - EU<br>SZ128913 | GGGG | ■ | FEATURED |
| ~ | Southbourne<br>SZ147911 | GG~~ | ■ | Backed by low sandstone cliffs. |
| G | **Hengistbury Head** - EU<br>SZ170906 | GGPG | ■ | FEATURED |
| | **Christchurch** | | | |
| G | Mudeford Sandbank - EU<br>SZ183912 | PPPG | ■ | FEATURED |
| ~ | Mudeford Quay<br>SZ183915 | FFP~ | ■ | |
| P | Avon Beach - EU<br>SZ183912 | PPPP | ■ | |

Sand · Shingle · Pebbles · Rocks · Mud · No information supplied

# South-West England

| RATING | NAME | TRACK RECORD | SEWAGE OUTLET | REMARKS |
|---|---|---|---|---|
| F | Friars Cliff - EU SZ192252 | PPPF | 🟩 | 🟨 🗑 |
| P | Highcliffe Castle - EU SZ200929 | PPPP | 🟩 | 🟨 |
| G | Highcliffe - EU SZ216931 | PP~G | 🟩 | 🟧 🟫 [??] |
| | **ISLE OF WIGHT** | | | |
| G | Totland - EU SZ322871 | PPFG | Screened, 300, 300m below LWM. | ⬆ 🟧 Not featured due to adjacent discharge. |
| G | Colwell Bay - EU SZ328879 | PPGG | (See Totland) | 🟧 Not featured due to adjacent discharge. |
| ~ | Yarmouth | ~~~~ | (See Norton) | 🟧 |
| G | Norton SZ347898 | P~~G | 1, 10,000, 230m below LWM. Preliminary treatment only. | ⬆ 🟨 Not featured due to adjacent discharge. |
| G | Gurnard Bay - EU SZ477959 | PPPG | 5,800, 400m below LWM. Preliminary treatment only. | 🟧 Not featured due to adjacent discharge. |
| P | **Cowes** West - EU SZ488967 | PPPP | 1, 15,000, LSO (700m). Preliminary treatment only. | ⬆ |
| P | East SZ506964 | PPPP | (see Cowes - West) | Sailing centre. |
| P | **Ryde** West SZ588930 | PFPP | 1, 21,000, 2km LSO. | 🟨 |
| F | East - EU SZ601927 | PPPF | (see Ryde West) | 🟨 |
| ~ | Seaview | ~~~~ | (see Ryde West) | 🟨 ⛰ |
| G | Bembridge - EU SZ657881 | PPPG | 1, 7,000, at LWM. | ⬆ 1998 Sailing centre. Not featured due to adjacent discharge. |
| P | Whitecliff Bay - EU SZ641862 | PPPP | (see Bembridge) | 🟨 |
| P | **Sandown** Yaverland SZ610849 | PFFP | Primary, 40,000, 250m below LWM. | ⬆ 1998 🟨 |
| P | Esplanade - EU SZ601842 | PPPP | (see Sandown Yaverland) | 🟨 |
| F | **Shanklin (Welcome Beach)** SZ589827 | FFPF | 🟩 | |
| P | Shanklin - EU SZ585811 | PPPP | 🟩 | 🟨 |
| P | Ventnor - EU SZ562773 | FFFP | 3, macerated, 11,500 100m offshore. | ⬆ 1998 🟨 |
| G | Compton Bay | GPPG | 🟩 | [??] 🟧 🗑 |
| G | St Helens - EU SZ637892 | PPPG | 🟩 | FEATURED |
| P | Seagrove Bay - EU SZ632912 | FPPP | 🟩 | 🗑 |

🟩 No discharge identified  ⬆ Improvements planned  [??] Insufficient information to feature  🗑 Cleaned regularly

*Hayling Island (p90) is representative of the South Coast's best beaches.*

# South-East England

The south-east coast of England is rich in contrasts: the area between Barton-on-Sea in Hampshire and Heacham in Norfolk encompasses some of the most varied and beautiful coastal scenery in the country, ranging from long shingle banks to sand dunes and saltmarshes.

•

Low clay cliffs predominate in Suffolk, while creeks and mud flats are found on the Essex coast and the Thames Estuary. Where the chalk hills of the South Downs reach the coast they form the formidable white cliffs of Dover, one of the most potent of national symbols and an impressive setting for many busy holiday resorts. Then there is Lowestoft Ness in Norfolk, the most easterly point of Britain, continually being eroded by the North Sea. Much of the Norfolk coast is designated an Area of Outstanding Natural Beauty and is managed to promote sustainable tourism.

The Solent is busy with boating traffic and the naval base of Portsmouth is a storehouse of British maritime history, home to famous ships such as the Mary Rose, HMS *Warrior* and HMS *Victory*. There are other ports of a very different nature – the ferry terminals of Felixstowe, Dover and Folkestone provide a gateway to Europe, while the Thames Estuary reminds us of our rich maritime past and is the site of popular boating areas such as the Medway, the Blackwater Estuary and the Swale.

These are some of our most heavily developed and densely populated coastal stretches and not without severe problems. Heavy shipping traffic in the Channel means that marine litter and oil are constantly being washed up on the beaches. International legislation is in place to prevent such pollution, but the signs are that it is being widely ignored. Throughout 1997 the Marine Conservation Society will be lobbying hard as important legislation on pollution by shipping passes through the House of Commons. Extensive stretches of the coastline suffer from pollution by sewage, although action is being taken at many sites in the south. One source of this pollution, the dumping of sewage sludge off the Thames Estuary, should end by 1998, but other discharges – domestic, nuclear and industrial – are still a cause for concern. The disturbance caused around Shakespeare Cliff during the construction of the Channel Tunnel has had, and will continue to have, serious long term effects on marine and coastal life. On a coastline under such pressure from diverse human activities including tourism, industry and residential development, great efforts are needed to bring about an improvement in the coastal environment and to ensure that those areas which have survived unspoilt are allowed to remain so.

# MILFORD-ON-SEA
## Hampshire
*OS Ref: SZ290913*

This is part of a long sweeping pebble beach interspersed with wooden and rock groynes, backed by low cliffs at its western end and adjoining a nature reserve to the east.

### Water Quality
No routine sewage discharge has been identified.

### Bathing Safety
Generally safe. There is no lifeguard cover, but safety signs operate.

### Litter
Dogs are allowed on this beach, which is cleaned daily in the summer.

### Access
Signposted from the A337. There are various entrances to the beach from the local road, stepped in the west and ramped in the east; disabled access is limited.

### Parking
Several large car parks nearby.

### Public Transport
Served by bus from Bournemouth and Southampton.

### Toilets
These include facilities for disabled visitors.

### Food
Cafés in two of the car parks and facilities in Milford village itself.

### Seaside Activities
Swimming.

*Beach huts nestle below the cliffs at Milford.*

### Wet Weather Alternatives
Limited locally but there are a number of New Forest attractions within reach.

### Wildlife and Walks
The Solent Way passes by here and many local routes exist; contact the Tourist Information Centre for further details.

### Tourist Information
Lymington Visitors Information Centre, New Street, Lymington, Hampshire.

### Track Record PGPG
EU designated beach.

# LEPE
## Hampshire
*OS Ref: SZ456385*

This sand and shingle beach, part of Lepe Country Park, is located on the Solent shore in an Area of Outstanding Natural Beauty, and looks across to the Isle of Wight.

**Water Quality**
No routine sewage discharge has been identified.

**Bathing Safety**
Generally safe; an inshore rescue service operates at weekends in the summer.

**Litter**
Park staff clean the beach daily. Bins are provided and dogs are banned from part of the beach during the summer.

**Access**
Take the A326 from Southampton to Holbury. Follow signs for Exbury, then Lepe. The beach is reached directly from the car park.

**Parking**
There are parking facilities.

**Public Transport**
Solent Blue Line buses serve the beach on summer Sundays.

**Toilets**
Toilets include facilities for disabled visitors.

**Food**
There is a restaurant, kiosk and ice cream van.

**Seaside Activities**
Swimming, sunbathing, windsurfing and fishing.

**Wet Weather Alternatives**
Local attractions in New Forest Area.

**Wildlife and Walks**
The Park runs a guided walks programme.

**Tourist Information**
Lyndhurst Tourist Information Centre, Lyndhurst, Hants.

**Track Record GPGG**
EU designated beach.

# HILL HEAD
## Hampshire
*OS Ref: SU548180*

A rural beach essentially of shingle with a lower foreshore of mud, sand and scattered deposits of shingle. There is an open flat grassed area to the rear of the beach lying between the car parks at Monks Hill and Crofton Lane.

**Water Quality**
One outfall serving 245,000 people discharges secondary treated effluent one kilometre below LWM.

**Bathing Safety**
Generally safe; there is no lifeguard cover.

**Litter**
Litter bins are located along the beach, which is cleaned daily during the summer, and in the car parks. Dogs are banned between May and September, and there is a poop-scoop scheme is in operation on the grassed area and promenade.

**Access**
Take the A27 from Fareham town centre to Peak Lane for Stubbington, Crofton Lane to Salterns car park, or Stubbington Lane to Monk's Hill car park. A path leads from the car parks to the promenade and beach.

**Parking**
There are spaces for 133 vehicles at Salterns car park and a further 100 at Monk's Hill.

**Toilets**
These include facilities for disabled visitors. Baby changing facilities exist at Salterns car park.

**Food**
There is a small beach café at Monk's Hill.

**Seaside Activities**
Swimming, sailing, windsurfing and fishing.

**Wet Weather Alternatives**
Adjacent to the beach is Titchfield Haven Nature Reserve.

**Wildlife and Walks**
The Solent Way runs along this stretch of the coast.

**Tourist Information**
Westbury Manor Museum, West Street, Fareham, Hants, PO16 0FF. Tel. 01329 824895

**Track Record GPGG**
EU designated beach.

# HAYLING ISLAND (EAST AND WEST BEACH)
## Hampshire
*OS Refs: SZ729984, SZ705987*

Eight kilometres of pebble beach with sand at low tide stretches along Hayling Island, from Eastoke Point at the entrance of Chichester Harbour to Sinah Common at the entrance of Langstone harbour. The western end of the beach is undeveloped, backed by dunes and a golf course.

**Water Quality**
No routine sewage discharge has been identified.

**Bathing Safety**
Unsafe at either end of the beach due to harbour currents. The Coastguard is on the foreshore.

**Litter**
Litter is hand picked twice daily during the season. Dogs are banned from May to September and by-laws prohibit fouling by dogs along the foreshore.

**Access**
Leave the A27 at its junction with the A3023, follow signs for Hayling Island.

**Parking**
There is plenty of parking on the foreshore.

**Public Transport**
A number of buses run from Havant which is also the nearest railway station.

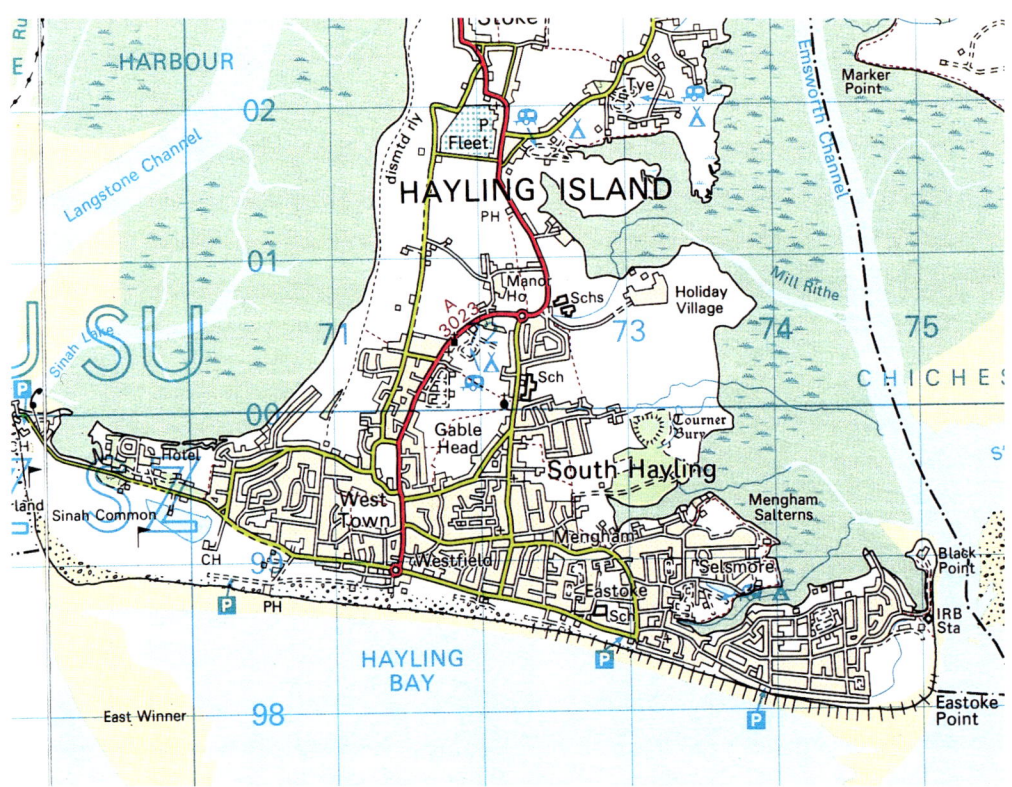

*South-East England*

*Windsurfers and bathers share the west beach at Hayling Island.*

**Toilets**
Public toilets include facilities for the disabled visitor.

**Food**
There are several kiosks and cafés.

**Seaside Activities**
Swimming, sailing, windsurfing and water-skiing; golf course nearby.

**Wet Weather Alternatives**
Amusement arcades.

**Wildlife and Walks**
In spite of its popularity with visitors, many areas of the island are still unspoilt. The walk along the Hayling Billy Coastal Path, built on the dismantled railway line to the north of the town, takes you along the east shore of Langstone Harbour and leads to a nature reserve covering an area of marshland with a variety of plant and animal life.

**Tourist Information**
Hayling Island Tourist Information Centre, Central Beachlands, Seafront, Hayling Island, PO11 OAG.
Tel. 01705 467111

**Track Record**
**Hayling Island West GGGG**
**Hayling Island East P~GG**
Both beaches are EU designated.

# EAST AND WEST WITTERING
## West Sussex
*OS Refs: SZ8096, SZ768980*

The mile-long, dune-fringed sandy beach at West Wittering is reached by a signposted drive from the main street. The East Wittering beach is backed by grassland; a large expanse of sand is revealed at low tide.

**Water Quality**
No routine sewage discharge has been identified.

**Bathing Safety**
West Wittering has a beach patrol and safety patrol. The District Council operates a patrol boat at East Wittering.

**Litter**
The beaches are cleaned regularly during the summer and bins are provided at both sites. Dogs are banned from May to September on sections of the beaches.

**Access**
Take the A286 from Chichester and follow the signs for East or West Wittering.

**Parking**
There are 386 spaces at Marine Drive, East Wittering and 4500 spaces at West Wittering.

**Public Transport**
Buses run from Chichester.

**Toilets**
There are public toilets at both beaches which include facilities for the disabled visitor.

**Food**
There are kiosks and café facilities at the beaches.

**Seaside Activities**
Swimming, surfing, windsurfing.

**Wet Weather Alternatives**
Earnley Butterflies and Gardens.

**Tourist Information**
Chichester Tourist Information Centre, 29a South Street, Chichester, West Sussex, PO19 1AH. Tel. 01243 539434

**Track Record**
**West Wittering PPPG** - EU designated.
**East Wittering GG~G** - Not EU designated.

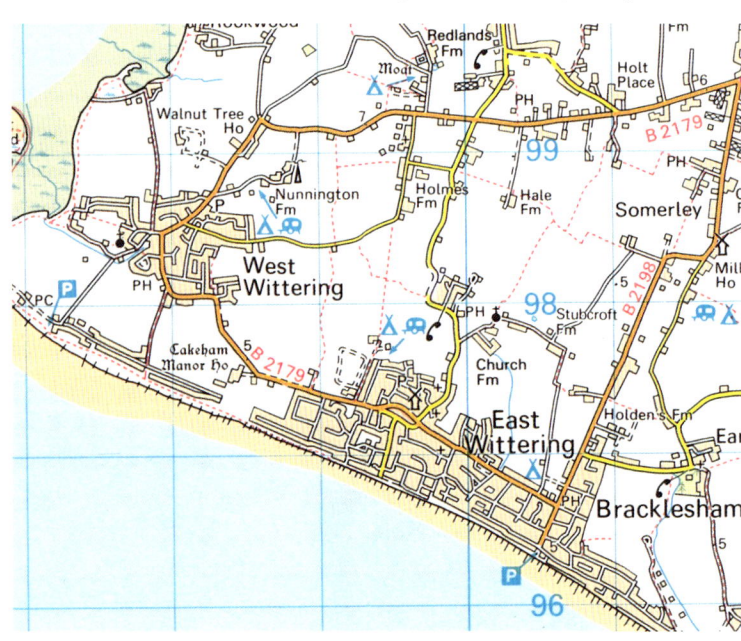

# SELEY BILL
## West Sussex
### OS Ref: SZ868937

From the low headland of Selsey Bill with its excellent views of the Isle of Wight, the beach extends in both directions. The groyne-ribbed east beach is the most popular, backed by the sea wall, with shingle stretching to Pagham Harbour. The Lifeboat Station and the fisherman's compound add to the interest. The west beach is backed by private land and stretches away to East Wittering round Bracklesham Bay.

**Water Quality**
No routine sewage discharge has been identified.

**Bathing Safety**
Bathing is safe one kilometre to either side of the headland, as fast currents around Selsey Bill makes bathing unsafe.

**Access**
There is direct access to the beach from the sea wall to the east of Selsey Bill.

**Parking**
A total of 500 spaces are available.

**Toilets**
Toilets are available at Selsey.

**Food**
A kiosk at East Beach opens during the bathing season; a pub near the lifeboat station sells food; various cafés can be found in the High Street.

**Seaside Activities**
Swimming, sailing and windsurfing.

**Wet Weather Alternatives**
Adjacent to the beach is Titchfield Haven Nature Reserve.

**Wildlife and Walks**
Pagham Harbour to the east is a nature reserve and a refuge for dozens of species of birds, butterflies and plants. The Sidlesham Ferry Nature Trail starting at the Visitors' Centre circles the western edge of the harbour and is well worth following to Pagham. The saltmarsh has a large number of visiting wildfowl in the winter.

**Tourist Information**
See Wittering (opposite).

**Track Record GGGG**
EU designated beach.

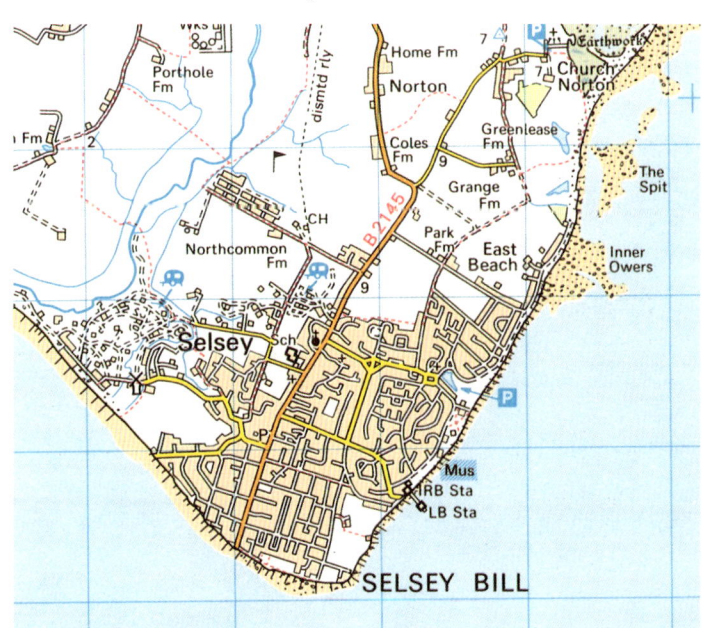

# RAMSGATE SANDS, RAMSGATE
## Kent
*OS Ref: TR387649*

This beach is a mix of sand and shingle, with safe bathing water adjacent to an historic and picturesque harbour and marina. The beach and the surrounding promenades have a wide range of family amusements and catering establishments.

### Water Quality
No routine sewage discharge has been identified.

### Bathing Safety
A designated area is patrolled by lifeguards from 10am to 6pm during the summer season.

### Litter
Cleaned regularly by hand and machine; litter bins are provided in strategic positions. Dogs are banned from the beach between May and September and must be kept on leads on the promenade.

### Access
Well signposted from the M2, the A299 and the A28. Ramps and steps lead from the promenade to the beach. Access for disabled visitors is good.

### Parking
Approximately 80 parking bays along the seafront.

### Public Transport
Frequent main-line rail links from London and a national coach service to Ramsgate. A local bus service stops within walking distance of the beach.

### Toilets
Toilets include baby changing and facilities for disabled visitors.

### Food
Seafront cafés, kiosks and restaurants.

### Seaside Activities
Swimming, boat rides, children's rides, trampolines and harbour walks.

### Wet Weather Alternatives
Various museums, amusement arcades, shopping, swimming pool, theatre, cinema.

### Wildlife and Walks
Smuggler's Walk, Ramsgate Harbour Trail, Pegwell Bay Nature Reserve and Thannet Coastal Walk.

### Tourist Information
Ramsgate Tourist Information Centre, 19/21 Harbour Street, Ramsgate, Kent, CT11 8HA. Tel. 01843 583353

**Track Record PPGG**
Not EU designated.

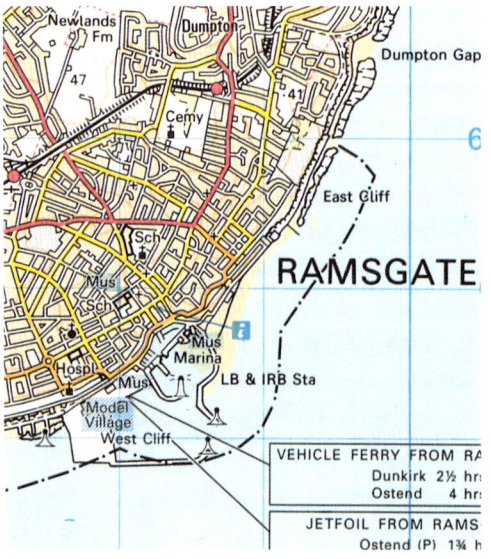

*South-East England*

# JOSS BAY, BROADSTAIRS
## Kent
*OS Ref: TR399702*

A natural sandy bay backed by low cliffs, this is a centre for canoeing and surfboarding, and there is good swimming too. Charles Dickens wrote his novel David Copperfield in this area, and the twisting narrow streets of Broadstairs are full of reminders: every year in June there is a week long Dickens Festival. A variety of bays and beaches abound providing widely differing characteristics and opportunities for recreation.

**Water Quality**
No routine sewage discharge has been identified.

**Bathing Safety**
Swimming is safe, with lifeguards in the summer season.

**Litter**
The beach is cleaned regularly by hand.

**Access**
Via the M2 Thanet Way to Broadstairs, then follow roadsigns.

**Parking**
There is a large car park adjacent to the beach.

**Toilets**
Toilets are available.

**Food**
A café and kiosk serve hot and cold snacks.

**Seaside Activities**
Swimming, surf ski hire and volleyball.

**Wet Weather Alternatives**
Local museums; theme park in Margate; day trips to France and Belgium via Ramsgate.

**Wildlife and Walks**
The well managed coastal nature trail is reached by following the signs from either end.

**Tourist Information**
Broadstairs Tourist Information Centre,
68 High Street,
Broadstairs,
Kent, CT10 1LH.
Tel. 01843 862242

**Track Record**
**GGGG**
EU designated beach.

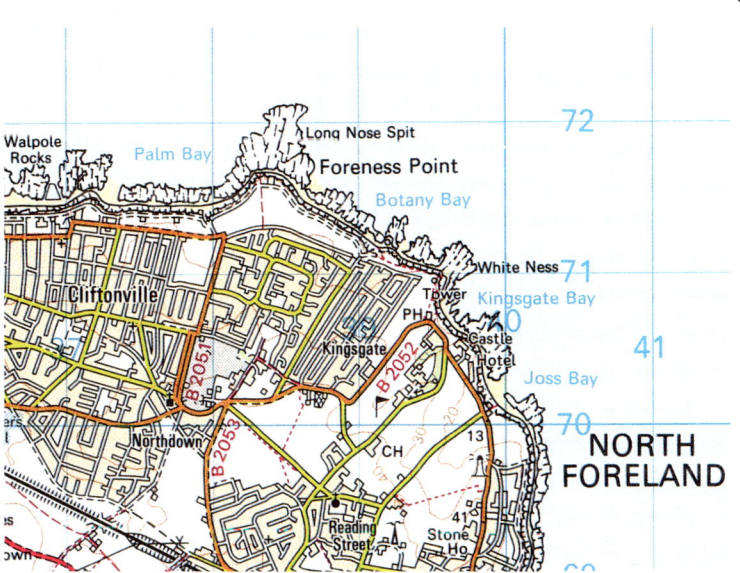

95

South-East England

# WESTGATE BAY (WEST BAY)
## Kent
*OS Ref: TR320702*

Craggy cliffs dip sharply on either side of this tide-washed sand-and-shingle beach. The gentle slope of the beach gives safe bathing conditions in most weathers.

**Water Quality**
No routine sewage discharge has been identified.

**Bathing Safety**
A designated area is patrolled by lifeguards from 10am to 6pm in summer.

**Litter**
The beach is cleaned by hand three times a day. Litter bins are provided in strategic positions. A poop-scoop scheme is in operation all year round.

**Access**
Follow signs from the M2, the A299 and the A28 to Birchington, then to Westgate. Ramps from the promenade give good access to the beach.

**Parking**
There is roadside parking with no restrictions.

**Public Transport**
Rail link to Westgate and a national coach service to Ramsgate. Local buses stop within walking distance of the beach.

**Toilets**
Includes facilities for disabled visitors.

**Food**
There is a café and kiosk.

**Seaside Activities**
Swimming, fishing. There are facilities for launching small craft.

**Wet Weather Alternatives**
Cinema, leisure centre, indoor swimming pool and squash courts (open to the public), Quex House museum (and see Ramsgate).

**Wildlife and Walks**
Thanet Coastal Walk.

**Tourist Information**
Margate Tourist Information Centre, 22 High Street, Margate, Kent, CT9 1DS. Tel. 01843 220241

**Track Record PPGG**
Not EU designated.

*The promenade is a popular location for sunseekers.*

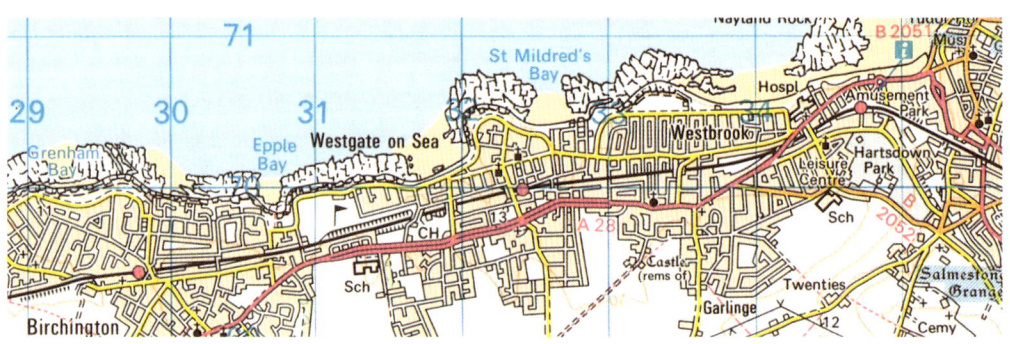

*South-East England*

# LEYSDOWN-ON-SEA (GROVE AVENUE)
## Kent
*OS Ref: TR034708*

Situated on the Isle of Sheppey, Leysdown-on-Sea has a sandy beach bounded by essential flood defences and backed by a promenade.

### Water Quality
No routine sewage discharge has been identified.

### Bathing Safety
Lifeguards on duty on the promenade from April to September. First aid is also available at the Beach Services Building. Perry buoys are fixed to the promenade.

### Litter
The beach is cleaned daily by hand during the bathing season and weekly during the winter. A dog ban is enforced from May until the end of September. Dogs must be kept on a lead on the promenade.

### Access
Access to the beach is by foot only, a walk of some 100m from the nearest car park and 200m from the bus stop. Steps and a specially constructed ramp make access suitable for disabled visitors.

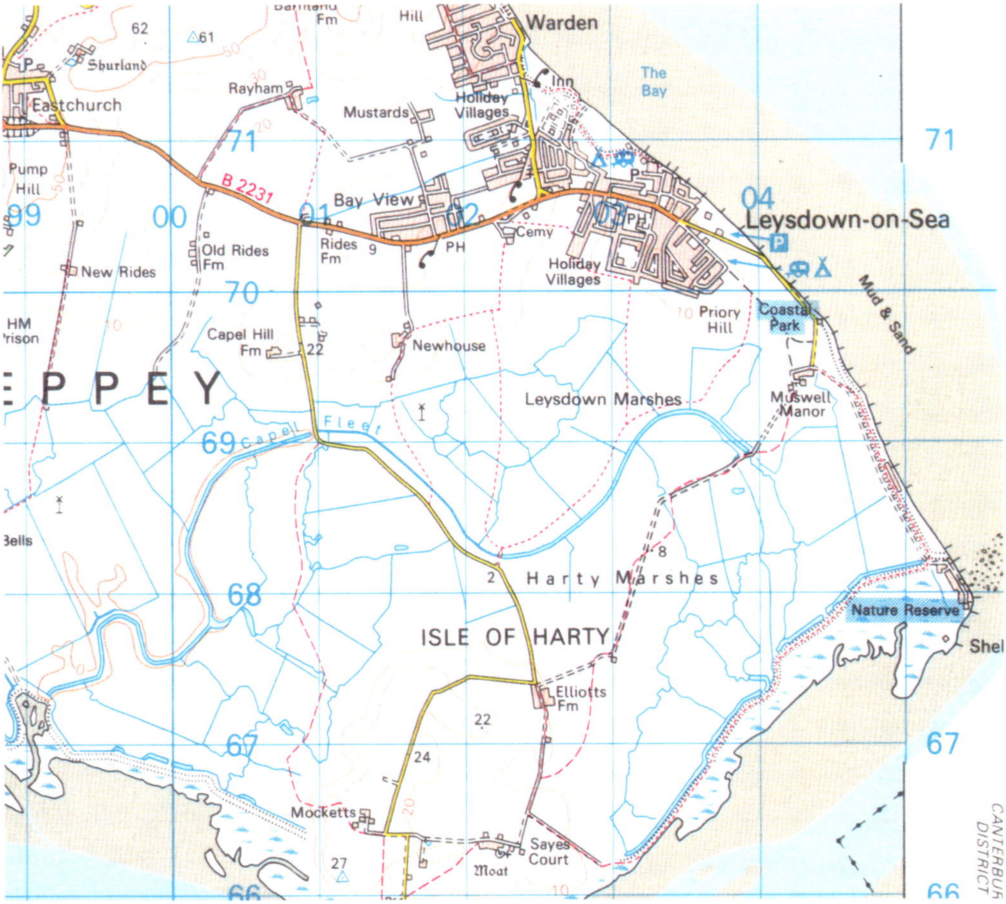

South-East England

## Parking
There is a car park on the promenade, 100m from the beach, with spaces for 250 cars.

## Public Transport
The nearest main line railway station is located at Sheerness. A regular bus service operates from Sheerness bus station to the beach.

## Toilets
These include facilities for the disabled and baby changing.

*Leysdown is one of Kent's most popular seaside resorts.*

## Food
Various cafés and pubs nearby.

## Seaside Activities
Swimming, windsurfing. Boating and sailing are available nearby.

## Wet Weather Alternatives
See Sheerness (overleaf).

## Wildlife and Walks
See Sheerness (overleaf).

## Tourist Information
See Sheerness (overleaf).

**Track Record PPGG**
EU designated beach.

South-East England

# SHEERNESS (BEACH STREET)
## Kent
*OS Ref: TR925750*

A substantial sea defence wall which forms the promenade is a distinctive characteristic of this gently sloping, 600 metre-long sand-and-shingle beach at the mouth of the Thames and Medway Estuaries. The beach is separated from the town centre by gardens, a leisure complex, swimming pool and a large sand-pit play area, and is a bathing-only zone.

### Water Quality
No routine sewage discharge has been identified.

### Bathing Safety
There is a lifeguard patrol between May and September, 9am to 5pm seven days a week, and first aid is available in the leisure complex and swimming pool. A flag system operates to warn of dangerous conditions; swimmers should keep away from the groynes.

### Litter
The beach is cleaned daily during the summer season. Dogs are banned from the beach between May and September.

### Access
From the M20, follow the signs for the A249 into the centre of Sheerness; the beach is signposted from here. Steps lead down from the promenade to the beach and lifeguards are happy to assist disabled visitors.

### Parking
There are 236 spaces in three car parks adjacent to the beach.

### Public Transport
Sheerness railway station is 150 metres from the beach. Buses arrive six times a day to the Tesco site, also 150 metres from the beach.

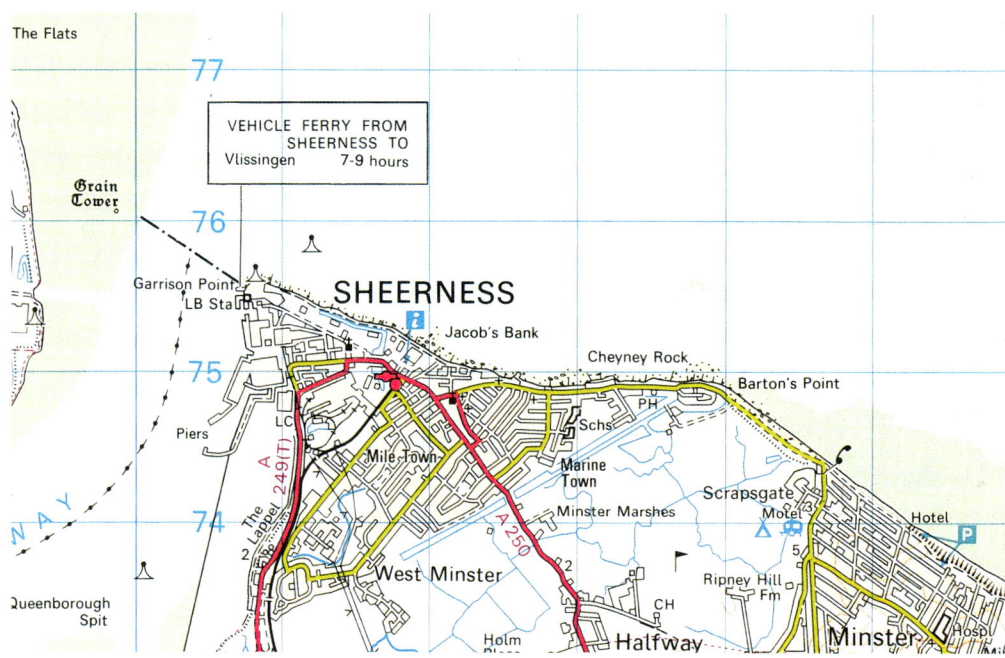

100

*South-East England*

*The massive sea wall above Sheerness Beach is a superb vantage point for views across the Thames Estuary.*

### Toilets
There are toilets in the leisure complex, at the swimming pool and on the street, with facilities for the disabled.

### Food
There is a wide range of cafés and restaurants in the high street, and a beach kiosk serves snacks.

### Seaside Activities
Bathing, paddling, sand-pit, climbing frame, sailing and windsurfing.

### Wet Weather Alternatives
The leisure complex and indoor heated pool, Minster Abbey Gatehouse, crazy golf, tennis and children's play ground.

### Wildlife and Walks
From Sheerness it is possible to walk right around the Isle of Sheppey, or walk one of the two coastal trails. The nearby Minster Cliffs are an outstanding site for a relaxing picnic. The Medway Estuary and Swale marshes are rich in wildlife and naturalists should visit the Swale National Nature Reserve at the eastern tip of Sheppey or the RSPB's Elmley Reserve. Turn right one kilometre from Kingsferry Bridge on to the A249. Once visitors are off the road they will be surprised by the wilderness of Southern Sheppey. The area is internationally important for wading bird populations.

### Tourist Information
Sheerness Tourist Information Centre, Sheppey Leisure Centre, Seafront, Sheerness. Tel. 01795 668061

**Track Record GGGG**
EU designated beach.

*South-East England*

# SHOEBURYNESS EAST
## Essex
*OS Ref: TQ945852*

This small sand and shingle beach backed by a grass field is popular in the summer. Rampart Terrace, which runs behind the beach, has fine views of the Thames estuary.

**Water Quality**
No routine sewage discharge has been identified.

**Bathing Safety**
Swimming is generally safe.

**Litter**
The beach is cleaned daily during the summer; dogs are banned between May and September and by-laws prohibit fouling the grassy area at all times. Bins are located in the car park and on the beach.

**Access**
From Southend Seafront to Thorpe Esplanade then Ness Road. Access to the beach is via the grassy headland.

**Parking**
Around 1,000 spaces in the car parks.

**Public Transport**
Shoeburyness railway station is five minutes walk from the beach.

**Toilets**
These include facilities for disabled visitors.

**Food**
There is a café nearby.

**Seaside Activities**
Swimming and windsurfing.

**Wet Weather Alternatives**
There are many attractions at Southend, approximately 3 miles away.

**Wildlife and Walks**
The area is of international importance for migrating birds.

**Tourist Information**
Southend Tourist Information Centre, 19 High Street, Southend. Tel. 01702 215120

**Track Record PPPG**
EU designated beach.

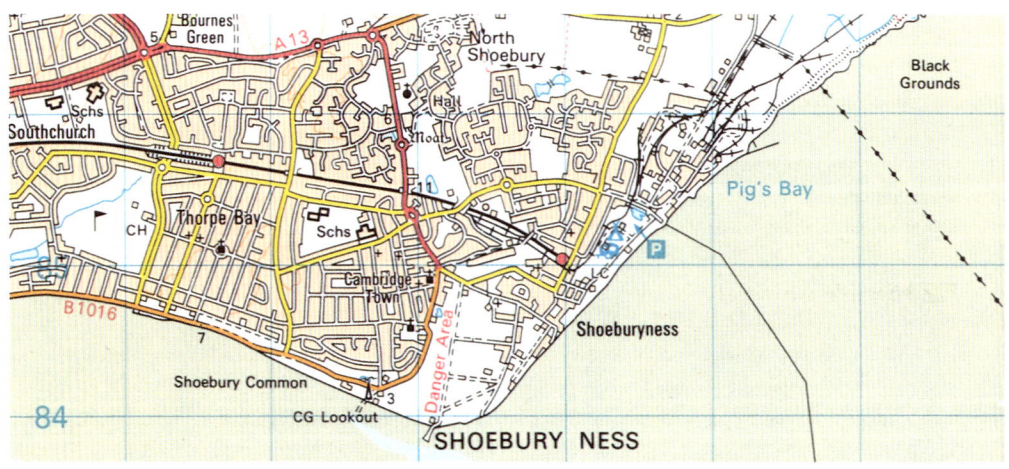

## BRIGHTLINGSEA
### Essex
*OS Ref: TM076161*

This small, semi-rural, enclosed beach has a tidal paddling pool which is very popular with children. Brightlingsea, once an important boat-building centre, is, surrounded on three sides by water, and can only be reached by a single road.

### Water Quality
One outfall serving 9,000 people discharges secondary treated effluent to the River Colne.

### Bathing Safety
Bathing is generally safe and lifebelt stations are present.

### Litter
The beach is cleaned daily in the summer; bins are only present during the season. Dogs are banned from the beach between May and September.

### Access
Follow the B1029 to Brightlingsea. Footpaths lead to the beach, and there is good access for disabled visitors.

### Parking
Space for 100 vehicles.

### Public Transport
There is a bus from Colchester.

### Toilets
Therse include facilities for disabled visitors.

### Food
A café serves hot and cold snacks.

### Seaside Activities
Bathing, paddling pool, outdoor swimming pool and boating lake.

### Wildlife and Walks
There is a local nature walk to the Brightlingsea Marsh National Nature Reserve.

### Tourist Information
Clacton Tourist Information Centre, Pier Avenue, Clacton-on-Sea, Essex. Tel. 01255 423400

**Track Record GGPG**
EU designated beach.

# CROMER
## Norfolk
*OS Ref: TG219425*

Popular as a resort since the 18th century, Cromer stands on a low crumbling cliff facing the North Sea. Winding streets lead past old flint cottages to the promenade, from where a slipway gives access to a long sand and shingle beach.

### Water Quality
No routine sewage discharge has been identified.

### Bathing Safety
Generally safe; lifeguards patrol between 10am and 6pm during the season.

### Litter
The beach is cleaned daily in the summer season. Dogs are banned from the beach between May and September, and a poop-scoop scheme is in operation.

### Access
On the A148 from Kings Lynn or the A140 from Norwich. There are car parking facilities for disabled visitors on the promenade; access is by slope and steps.

### Parking
Approximately 50 spaces on the promenade for disabled visitors.

### Public Transport
Train and bus services run to Cromer from Norwich, and local services include a coastal route.

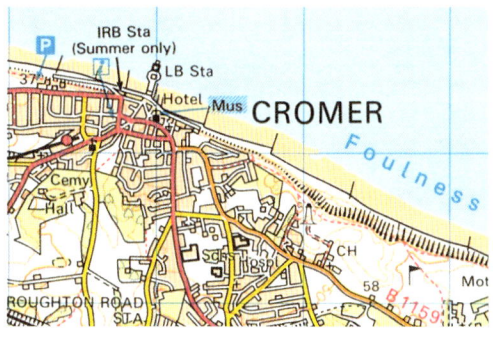

### Toilets
These include baby changing and facilities for disabled visitors.

### Food
There are cafés and pubs nearby.

### Seaside Activities
Swimming, surfing, fishing from the pier and beach. A beach club operates during school summer holidays.

### Wet Weather Alternatives
End of pier family show, amusement arcades on promenade and children's fun fair. National Trust properties close by.

### Wildlife and Walks
There is a coastal path and Holt Country Park is adjacent to the beach.

### Tourist Information
Cromer Tourist Information Centre, Cromer Bus Station, Town Centre, Cromer. Tel. 01263 512497

**Track Record PFPG**
EU designated beach.

*South-East England*

# SHERINGHAM
## Norfolk
*OS Ref: TG162436*

The coming of the railway in 1887 turned Sheringham into a popular summer destination. Low cliffs form an idyllic backdrop to the shingle and sand beach which is augmented at low tide by a vast expanse of sand.

### Water Quality
No routine sewage discharge has been identified.

### Bathing Safety
Generally safe; lifeguards patrol between 10am and 6pm during the season.

### Litter
The beach is cleaned daily in the summer season, and bins are provided. Dogs are banned from the beach between May and September, and a poop-scoop scheme is in operation.

### Access
Off the A149 Great Yarmouth to Kings Lynn road or the A148 Kings Lynn to Cromer road. Access ramps run from the high street to the traffic-free promenade.

### Parking
Parking is available on the cliffs but not on the promenade.

### Public Transport
The train to Sheringham can be caught from Norwich. Local bus services from all East Anglian towns, coastliner route.

### Toilets
These include facilities for disabled visitors.

### Food
There are promenade cafés, restaurants and hotels within 50 metres of the promenade.

### Seaside Activities
Swimming, windsurfing, surfing, fishing. There is a boat launching ramp, and a beach club operates during school summer holidays.

### Wet Weather Alternatives
Amusements in town centre; local history museum; North Norfolk Railway museum. Sheringham Park (National Trust) is close by.

### Wildlife and Walks
Sheringham is located in an Area of Outstanding Natural Beauty and the coastal footpath gives access to a Site of Special Scientific Interest. Contact the Tourist Information Centre for details.

### Tourist Information
Sheringham Tourist Information Centre, (next to town car park and steam railway), Sheringham.
Tel. 01263 824329

### Track Record GGPG
EU designated beach.

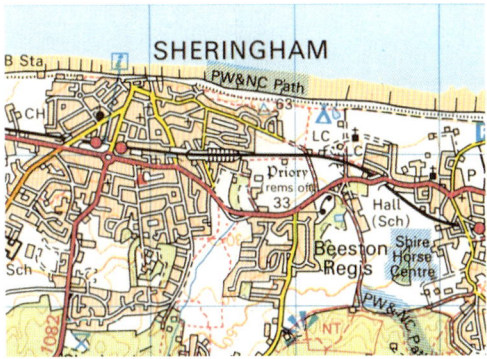

# HUNSTANTON
## Norfolk
*OS Ref: TF678425*

The only coastal town in East Anglia to face west, Hunstanton is known for its famous red-and-white striped cliffs. The tide retreats to expose a large sandy beach, which can become crowded in the summer.

### Water Quality
No routine sewage discharge has been identified.

### Bathing Safety
New first aid post, lost children's post, lifeguard cover and warning signs introduced for 1997.

### Litter
The beach is cleaned regularly. Dogs are banned from the beach during the summer, and a poop-scoop scheme is in operation. Bins are provided for both litter and dog waste.

### Access
Take the A149 to Hunstanton. Steps and ramps from the promenade lead to the beach.

### Parking
There are plenty of parking spaces in Hunstanton.

### Public Transport
A bus runs twice daily from Kings Lynn to Hunstanton.

### Toilets
These include facilities for disabled visitors.

### Food
There are a large number of cafés, restaurants and pubs in the vicinity.

### Seaside Activities
Swimming, windsurfing and sailing.

### Wet Weather Alternatives
Sealife centre, Oasis leisure centre, Wonderland soft play centre, Sandringham and Park Farm.

### Wildlife and Walks
Peddars Way and Norfolk Coastal Path; for further information contact the Tourist Information Centre.

### Tourist Information
Hunstanton Tourist Information Centre, Town Hall, The Green, Hunstanton. Tel. 01485 532610

### Track Record GPPG
EU designated beach.

*Summer draws visitors in their droves.*

*South-East England*

## South-East England

| RATING | NAME | TRACK RECORD | SEWAGE OUTLET | REMARKS |
|---|---|---|---|---|
| | HAMPSHIRE | | | |
| G | **Chistchurch Bay** (**Barton-on-Sea**) SZ238928 | P~~G | Primary and disinfected, 7,000, at LWM. | ⬆ 1997 🟨 🗑 Not featured due to adjacent discharge. |
| G | **Milford-on-Sea** - EU SZ290913 | PGPG | 🟩 | FEATURED |
| G | **Lepe** - EU SZ456385 | GPGG | 🟩 | FEATURED |
| G | **Calshot** - EU SU481012 | PPPG | 2, preliminary only, 16,000, at LWM. | ⬆ 🟫 Not featured due to adjacent discharge. |
| ~ | **Solent Breezes** SU505038 | P~~~ | 🟩 | |
| G | **Hill Head** - EU SU548180 | GPGG | (See Lee-on-the-Solent) | FEATURED |
| P | **Lee-on-the-Solent** - EU SU562005 | GPGP | Secondary, 245,000, 1,000m below LWM, from Peel Common. | 🟨 🟧 |
| F | **Stokes Bay, Gosport** - EU SZ600979 | GGPF | (See Lee-on-the-Solent) | 🟨 |
| G | **Old Portsmouth Beach** SZ631992 | ~~~G | 🟩 | 🟨 Not featured as unsuitable for bathing. |
| F | **Southsea** (South Parade Pier) - EU SZ653982 | FFFF | (See Eastney) | 🟧 |
| G | **Eastney** - EU SZ675988 | GPPG | 1, 240,000, 5.7km, LSO | ⬆ 🟨 Not featured due to adjacent discharge. |
| G | **Hayling Island** West - EU SZ705987 | GGGG | 🟩 | FEATURED |
| G | East - EU SZ729984 | P~GG | 🟩 | FEATURED |
| | WEST SUSSEX | | | |
| G | **West Wittering** - EU SZ768980 | PPPG | 🟩 | FEATURED |
| G | **East Wittering** SZ8096 | GG~G | 🟩 | FEATURED |
| ~ | **Bracklesham Bay** - EU SZ805963 | ~PG~ | 🟩 | 🟨 🟧 |
| G | **Selsey Bill** - EU SZ868937 | GGGG | 🟩 | FEATURED |
| G | **Pagham** - EU SZ892972 | PPPG | (See Bognor Regis) | 🟧 Not featured due to adjacent discharge. |
| G | **Bognor Regis** - EU SZ923985 | PPFG | Screened, 71,500, 2.6km, LSO. | ⬆ 🟨 🗑 Not featured due to adjacent discharge. |
| G | **Bognor Regis - Pier** - EU SZ937988 | ~~~G | | 🟨 Not featured due to adjacent discharge. |
| P | **Felpham** (Yacht Club) - EU SZ985993 | P~~P | 🟩 | 🟨 |
| G | **Middleton-on-Sea** - EU SZ985999 | PPGG | | ❓ 🟨 🟧 |
| P | **Littlehampton** - EU TQ040013 | PPPP | Screened, 60,000, LSO. | ⬆ 🟨 🟧 🗑 |

108  🟨 Sand    🟧 Shingle    🟦 Pebbles    ⛰ Rocks    🟫 Mud    ❓ No information supplied

## South-East England

| RATING | NAME | TRACK RECORD | SEWAGE OUTLET | REMARKS |
|---|---|---|---|---|
| ~ | **Goring-by-the-Sea** | ~~~~ | ■ | 🟨 |
| P | **Worthing** West - EU TQ139021 | FFPP | ■ | 🟨 🟧 🗑 |
| ~ | East TQ168029 | F~~~ | Primary, 105,000, LSO. | 🟨 🟧 🗑 |
| G | **Lancing** (South) - EU TQ183036 | FFPG | (See Worthing East) | 🟨 🟧 Not featured due to adjacent discharge. |
| ~ | **Shoreham** (Kingston Beach) TQ235046 | P~~~ | ■ | 🟨 🟧 |
| ~ | **Shoreham-by-Sea Beach** TQ214047 | P~~~ | Primary, 6,000, 3.1km, LSO. | Commercial port. |
| G | **Southwick** - EU TQ242048 | PFPG | ■ | 🟧 ⁇ |
| | EAST SUSSEX | | | |
| G | **Hove** - EU TQ288043 | PFPG | Stormwater. | ⬆ 1997 🟧 Not featured due to adjacent discharge. |
| P | **Brighton** Kemp Town - EU TQ323034 | FFPP | Stormwater. | ⬆ 1997 🟨 🟧 |
| ~ | Palace Pier TQ314038 | F~~~ | Stormwater. | ⬆ 1997 🟨 🟧 |
| G | **Saltdean** - EU TQ381018 | PPGG | | 🟨 ⛰ 🗑 Not featured due to adjacent discharge. |
| ~ | **Portobello** | ~~~~ | Screened, 254,000, 1.5km LSO, plus stormwater. | ⬆ 2000 |
| F | **Newhaven** - EU TV449998 | P~PF | Screened, 35,000, 2.6km LSO. | ⬆ 2000 🟨 |
| G | **Newhaven West Quay** TV447999 | ~PGG | | 🟨 Within breakwater. Not featured due to adjacent discharge. |
| P | **Seaford** - EU TV488982 | GFGP | ■ | 🟧 🗑 |
| G | **Cuckmere Haven** TV520975 | PPGG | ■ | ⁇ 🖼 Bathing not safe near mouth of the river. |
| P | **Birling Gap** TV552060 | ~PGP | ■ | Affected by marine litter. |
| P | **Eastbourne** East of Pier TV625998 | P~PP | Primary 130,000, 3.2 km LSO. Also stormwater. | 🟨 🟧 |
| P | Wish Tower - EU TV614982 | PPPP | (see Eastbourne - East of Pier) | 🟨 🟧 🗑 |
| P | **Pevensey Bay** - EU TQ657037 | PPPP | ■ | 🟨 🟧 🗑 |
| P | **Norman's Bay** - EU TQ682053 | PPPP | ■ | 🟨 🟧 🗑 |
| ~ | **Cooden Beach** | F~~~ | ■ | 🟨 🟧 |
| P | **Bexhill** (Egerton Park) - EU TQ737068 | PPPP | Stormwater. | ⬆ 2000 🟨 🟧 🗑 |
| ~ | **Hastings** Bulverhythe TQ784086 | P~~~ | 2, preliminary only, 168,000 LSO. | ⬆ 🟧 |

■ No discharge identified  ⬆ Improvements planned  ⁇ Insufficient information to feature  🗑 Cleaned regularly

## South-East England

| RATING | NAME | TRACK RECORD | SEWAGE OUTLET | REMARKS |
|---|---|---|---|---|
| P | St Leonard's Beach TQ797087 | P~PP | (see Hastings - Bulverhythe) | |
| F | Queens Hotel - EU TQ819092 | FFPF | (see Hastings - Bulverhythe) | |
| ~ | Fairlight Glen TQ912154 | P~~~ | ■ | Rockpools. |
| P | Winchelsea - EU TQ912154 | GPGP | ■ | |
| P | Camber Sands - EU TQ973184 | PPPP | (see Broomhill Sands) | Sand dunes. |
| ~ | Broomhill Sands | ~~~~ | Secondary, 11,000 (inc. tourists), above LWM. | |
| | **KENT** | | | |
| ~ | Greatstone Beach TR082229 | P~~~ | (see Littlestone-on-Sea) | |
| P | Littlestone-on-Sea - EU TR084239 | PPPP | Secondary and UV, 8,000, between HWM and LWM. | |
| P | St Mary's Bay - EU TR093277 | PPPP | ■ | |
| | **Dymchurch** | | | |
| ~ | Hythe Road TR128319 | P~~~ | (see Dymchurch Beach) | |
| P | Dymchurch Beach - EU TR113304 | PPPP | Secondary and UV,10,000, between HWM and LWM. | Backed by sea wall. |
| ~ | Redoubt TR101290 | P~~~ | (see Dymchurch Beach) | |
| G | Hythe - EU TR160340 | PGGG | Preliminary only, 31,000, LSO. | Not featured due to adjacent discharge. |
| | **Sandgate** | | | |
| P | Sandgate Beach - EU TR188348 | PPGP | ■ | Steeply sloping. |
| ~ | Town Centre TR203351 | F~~~ | ■ | |
| F | Folkestone - EU TR237363 | PFPF | Raw, 68,500, at LWM. | 1999 |
| P | The Warren TR248376 | PPPP | (see Folkestone) | Danger from falling rocks. |
| ~ | Shakespeare Cliff | ~~~~ | Screened, 260,000, 400m offshore | 1999 |
| P | Dover Harbour TR321412 | PPGP | | Not featured due to adjacent discharge. |
| G | St Margaret's Bay - EU TR368444 | GPGG | ■ | ?? |
| G | Kingsdown Beach TR380485 | ~~~G | ■ | ?? |
| G | Deal Castle - EU TR378052 | FFGG | ■ | ?? |
| G | Sandwich Bay - EU TR358590 | FFFG | ■ | ?? |
| | **Ramsgate** | | | |
| G | Ramsgate Sands TR387649 | PPGG | ■ | FEATURED |

110 · Sand · Shingle · Pebbles ·  Rocks · Mud ·  No information supplied

## South-East England

| RATING | NAME | TRACK RECORD | SEWAGE OUTLET | REMARKS |
|---|---|---|---|---|
| P | Ramsgate Beach - EU TR372640 | FFPP | ■ | □ |
| P | **Broadstairs** East Cliff TR401688 | PPPP | 29,500, preliminary only, 3.6 km LSO. | □ |
| F | Broadstairs Beach - EU TR398677 | FPPF | (see Broadstairs, East Cliff) | □ |
| G | Joss Bay - EU TR399702 | GGGG | ■ | FEATURED |
| G | **Botany Bay**, Cliftonville TR391712 | PGGGG | 70,000, 1.8km, LSO at Foreness Point. | ↑ □ Not featured due to adjacent discharge. |
| G | **Palm Bay** TR373714 | PPPG | (see Botany Bay) | □ Not featured due to adjacent discharge. |
| P | **Walpole Bay** TR365715 | PPGP | ■ | □ ▲ |
| P | **Margate** The Bay - EU TR347708 | PPPP | ■ | □ |
| P | Fulsam Rock - EU TR356715 | GPGP | ■ | □ |
| P | **Westbrook Bay** TR320705 | PPPP | ■ | □ |
| P | **St Mildred's Bay** - EU TR328705 | GPPP | ■ | □ |
| G | **Westgate Bay** TR320702 | PPGG | ■ | FEATURED |
| P | **Minnis Bay** - EU TR286697 | PGGP | ■ | □ ▲ ♻ Safe for bathing. |
| P | **Herne Bay** - EU TR186686 | PFPP | ■ | □ ◨ |
| G | **Hampton Pier Beach** TR158684 | ~~~G | ■ | ⁇ □ |
| P | **Tankerton Beach** TR127674 | ~~~P | ■ | ▦ |
| P | **Whitstable** TR1166 | G~~P | 1, 32,000, preliminary only, 1.5km, LSO | ↑ ▦ Currents can make bathing dangerous. |
| G | **Leysdown-on-Sea** - EU TR034708 | PPGG | ■ | FEATURED |
| G | **Sheerness** (Beach Street) - EU TQ925750 | GGGG | ■ | FEATURED |
| ~ | **Minster Leas** | G~~~ | ■ | ♻ |
| P | **Allhallows** | ~~~P | | |
| | ESSEX | | | |
| P | **Canvey Island** TQ805824 | PPPP | Secondary, 44,500, at LWM. | □ ■ |
| P | **Leigh-on-Sea** | PFPP | ■ | □ ♻ |
| F | **Westcliff-on-Sea** - EU TQ864852 | PPPF | | ↑ □ ▦ ♻ |

■ No discharge identified   ↑ Improvements planned   ⁇ Insufficient information to feature   ♻ Cleaned regularly

## South-East England

| RATING | NAME | TRACK RECORD | SEWAGE OUTLET | REMARKS |
|---|---|---|---|---|
| P | Southend-on-Sea TQ887850 | PPPP | Primary, 198,000, 500m below LWM. | ⬆ 🟨 🟧 🟫 |
| P | Thorpe Bay - EU TQ911847 | PFPP | Secondary, 198,000, LSO | ⬆ 🟨 🟧 🗑 |
| P | Shoeburyness TQ925841 | PPPP | | ⬆ 1998 🟨 🟧 🗑 |
| G | Shoeburyness East - EU TQ945852 | PPPG | 🟩 | FEATURED |
| P | West Mersea - EU TM022120 | FFPP[ | Secondary and UV, 9,000. | 🟨 🟧 |
| G | Brightlingsea - EU TM076161 | GGPG | Secondary, 9,000, to River Colne. | FEATURED |
| P | Jaywick - EU TM148128 | PPPP | Screened, 24,000, LSO | ⬆ 2000 🟨 |
| P | Clacton  Off Coastguard Station TM173142 | P~~P | 🟩 | ⬆ 1998 🟨 |
| P | Groyne TM175144 | P~PP | 🟩 | ⬆ 1998 🟨 |
| G | Connaught Gardens - EU TM187152 | PPPG | 2, 300m and 50m below LWM, stormwater. | ⬆ 1998 🟨 Not featured due to adjacent discharge. |
| G | Holland-on-Sea - EU TM224176 | PPGG | Screened, 50,000, LSO. | ⬆ 2000 🟨 Not featured due to adjacent discharge. |
| P | Frinton-on-Sea - EU TM237194 | PPPP | 2, 50m below LWM, stormwater. | ⬆ 1998 🟨 |
| P | Walton-on-the-Naze - EU TM225215 | PPPP | Secondary, 31,000, 50m below LWM. | ⬆ 1998 🟨 |
| P | Dovercourt - EU TM251306 | PPPP | Secondary, 39,000, discharge to River Stour. | ⬆ 🟨 |
| P | Harwich (Sailing Club) TM263326 | P~~P | (see Dovercourt) | ⬆ |
| | SUFFOLK | | | |
| G | Felixstowe  South - EU TM297337 | PPPG | (see Felixstowe North) | ⬆ 🟨 🟧 Not featured due to adjacent discharge. |
| P |   North - EU TM305343 | GGGP | Secondary, discharged to Orwell Estuary. | 🟨 🟧 |
| ~ | Aldeburgh TM4757 | G~~~ | Macerated, 4,000, LSO. | 🟨 🗑 🟧 Steep ridges. |
| ~ | Dunwich TM4770 | G~~~ | 🟩 | 🪨 |
| P | Southwold, The Denes - EU TM508754 | GGGP | 🟩 | 🟨 🟧 🗑 |
| ~ | Kessingland TM536867 | G~~~ | Secondary, 5,000, to inland watercourse. | 🟧 |
| P | Lowestoft  South Beach - EU TM545917 | PGPP | 🟩 | 🟨 🗑 |
| P |   North Beach - EU TM553950 | PPPP | Screened, 140,000, LSO. | ⬆ 1998 🟨 |
| P | Gorleston Beach - EU TG532031 | FFPP | 🟩 | 🟨 |

🟨 Sand   🟧 Shingle   🪨 Pebbles   ⛰ Rocks   🟫 Mud   ❓ No information supplied

# South-East England

| RATING | NAME | TRACK RECORD | SEWAGE OUTLET | REMARKS |
|---|---|---|---|---|
| | **NORFOLK** | | | |
| | **Great Yarmouth** | | | |
| ~ | Power station | F~~~ | (see Great Yarmouth North) | 🟨 |
| P | South - EU TG533064 | FFFP | (see Great Yarmouth North) | 🟨 |
| P | Pier - EU TG533074 | FFFP | (see Great Yarmouth North) | 🟨 |
| P | North - EU TG535105 | PPPP | Many, screened, 75,000 (total), LSO | 🟨 |
| G | Caister Point - EU TG530120 | PPPG | Screened, 170,000, LSO | ⬆️ 🟨 2000 Not featured due to adjacent discharge. |
| G | Scatby Beach TG515155 | ~~~G | (see Great Yarmouth North) | ⁉️ 🟨 |
| G | Hemsby - EU TG509174 | PPGG | | ⁉️ |
| ~ | Sea Palling TG249412 | P~~~ | 🟩 | 🟨 〰️ Backed by sea wall and sand dunes. |
| ~ | Happisburgh | ~~~~ | 🟩 | 🟨 |
| P | Mundesley - EU TG317366 | PPPP | Screened, 7,000, LSO. | ⬆️ 2000 🟨 Backed by cliffs. |
| ~ | Overstrand TG249412 | F~~~ | 🟩 | 🟨 |
| G | Cromer - EU TG219425 | PFPG | 🟩 | FEATURED |
| ~ | East Runton | ~~~~ | 🟩 | 🟨 |
| ~ | West Runton | ~~~~ | Primary, 30,000, LSO. | 🟨 |
| G | Sheringham - EU TG162436 | GGPG | 🟩 | FEATURED |
| P | Wells-next-the-Sea - EU TF914456 | PPPP | Secondary and UV, 5,000, to inland drain. | 🟨 Backed by pine forests and sand dunes. |
| | **Hunstanton** | | | |
| P | North Beach Sailing Club TF672412 | P~PP | 🟩 | 🟨 |
| P | Boat Ramp TF667400 | P~PP | 🟩 | 🟨 |
| P | South Beach, Hunstanton Road TF660395 | P~PP | 🟩 | 🟨 |
| G | Hunstanton - EU TF678425 | GPPG | 🟩 | FEATURED |
| | **Heacham** | | | |
| P | North Beach - EU TF663375 | PPFP | (see South Beach - nr. river) | |
| P | South Beach TF659362 | P~FP | (see South Beach - nr. river) | |
| F | South Beach (near River) TF661368 | P~FF | Secondary, UV, 25,200, to Heacham River. | ⬆️ 1997 Gravel. |
| P | Snettisham Beach TF647335 | P~PP | 🟩 | Near RSPB reserve. |

🟩 No discharge identified   ⬆️ Improvements planned   ⁉️ Insufficient information to feature   〰️ Cleaned regularly

*Built on an outcrop of rock, Bamburgh Castle presides over the dramatic sweeping beaches of this historic coastline.*

# THE EAST COAST

FROM THE VAST, OPEN SKIES AND WATERY MARSHLANDS OF THE WASH TO THE WILD AND REMOTE STRETCHES OF BEACH ALONG THE BORDERLANDS OF NORTHUMBRIA, THIS SECTION OF BRITAIN'S COASTLINE ENCOMPASSES THE SHARPEST OF CONTRASTS, EVOKING DESCRIPTIONS SUCH AS FABULOUS, SPECTACULAR, MYSTERIOUS AND DRAMATIC.

•

Sadly, much less complimentary terms have also been used in connection with this area – polluted, industrialised, spoilt and scarred are just some which come to mind. Here industry extends to the shore, bringing with it steel mills, power stations, oil refineries and chemical works. The toxic waste they discharge pollutes our seas and coasts and damages our precious wildlife.

There are, though, many beautiful beaches on the east coast with excellent water quality, particularly in Lincolnshire and Humberside, and visitors to these counties are rewarded by some of the most dramatic cliff walks in the country. To the north, the empty borderlands are characterised by rugged cliffs and offshore islands, for many years outposts of the early Christian faith. St Aidan, sent to Holy Island to convert the people of Northumbria, built a small church and over the years it developed into a Benedictine monastery, the remains of which can be seen today. Dozens of ruined castles attest to centuries of conflict between the Scots and the English: the most northerly English town, Berwick-upon-Tweed, changed its allegiance no less than 13 times over 300 years of struggle.

Further south, beyond the industrialisation of the Tyne Estuary, are limestone cliffs of great conservation value; strange rock formations give way to the sweeping sandy beaches and fishing villages of Robin Hood's Bay and Filey. The Humber Estuary has vast expanses of saltmarsh and at the margins of the Wash seals can be seen basking off Gibraltar Point. The Wash itself is one of the most important areas of mudflats, sandbanks and saltmarsh in the United Kingdom and has international significance as a wetland with an annual population totalling over 200,000 resident and migrating birds. In the past, large areas of mudflats were lost to agriculture, but no further land reclamation has been carried out since the late seventies.

But the region is beset by many problems, not the least of which has been the offshore dumping of millions of tonnes of dredged spoil, sewage sludge and fly ash each year (although on the positive side, the dumping of fly ash and sewage sludge is due to end in 1998). Years of dumping coal-mining waste has blackened the beaches of Durham and Cleveland, although a Millennium Fund Project to clean this up has had some success. The closure of mines along the east coast may lead to an improvement in the state of the beaches, although ironically this may be to the detriment of groundwater, which risks contamination by leachates from the mines. Raw and partially treated sewage pollutes and contaminates many of the beaches and bathing waters: shellfish gathered from the Wash are required by law to be cleansed and steam-cooked before they can be sold. Few of the beaches in the region escape the problem of marine litter.

# SUTTON-ON-SEA
## Lincolnshire
*OS Ref:TF522821*

This groyne-ribbed beach offers a wide welcoming expanse of soft sand. Recent sensitive development of the resort's facilities, particularly the surrounding gardens and children's paddling pool, has given Sutton a welcome boost. The peaceful, uncommercialised air and unhurried pace of life encourages young families and senior citizens alike.

**Water Quality**
No routine sewage discharge has been identified.

**Bathing Safety**
Bathing is safe and suitable for children. The beach is patrolled regularly by lifeguards who are in radio contact with the base station and the coastguard.

**Litter**
The beach is cleaned and brushed daily. Both bins and recycling facilities exist along the promenade. Dogs are banned between May and September, and a poop-scoop scheme runs throughout the year.

**Access**
The resort is signposted from the A16 and the A1111. The beach is reached from the promenade via steps or ramps suitable for disabled visitors.

**Parking**
There are several car parks within close proximity of the beach.

**Public Transport**
National Express run coaches to the resort; there is a regular bus service between the railway station and the beach.

**Toilets**
Situated along the beach front, these include facilities for disabled visitors.

**Food**
Various cafés and restaurants serve refreshments close to the beach.

**Seaside Activities**
Swimming, paddling pools, bowls,

*The East Coast*

*The soft sand at Sutton-on-Sea is particularly popular with families.*

putting and donkey rides. Beach chalets, deck chairs and windbreaks are for hire. An 18-hole golf course hugs the shore to the south of the town.

**Wet Weather Alternatives**
A theatre offers entertainment for both children and adults. Nearby Mablethorpe has many attractions, including a cinema, an amusement park and the Animal Gardens and Seal Trust.

**Wildlife and Walks**
There are various local walks, details of which can be obtained from the Tourist Information Centres.

**Tourist Information**
*Seasonal:* Sutton-on-Sea Tourist Information Centre, Central Beach, Sutton-on-Sea. Tel. 01507 441373

*All Year Round:* Mablethorpe Tourist Information Centre, The Dunes Theatre, Central Promenade, Mablethorpe. Tel. 01507 472496

**Track Record PGPG**
EU designated beach.

# REIGHTON SANDS
## Yorkshire
*OS Ref: TA144763*

Reighton Sands is a rural beach with a large expanse of sand, making up the southern end of the 4-mile-long Filey Beach. Backed by slumping cliffs of boulder clay, this area is a firm favourite with families, walkers and naturalists.

**Water Quality**
No routine sewage discharge has been identified.

**Bathing Safety**
Bathing is safe with care. There are no lifeguards on site but general cover is provided by the lifeguards of Filey Beach on Land Rover patrol. An emergency telephone and lifebuoy are located on the beach.

**Litter**
There are no litter bins and the beach is cleaned when necessary. There is no dog ban on this stretch of sand.

**Access**
Follow the signs to Reighton Gap off the A165.

**Parking**
There is limited parking and no direct vehicle access to the beach.

**Public Transport**
A regular bus service runs between Scarborough and Filey.

**Toilets**
Toilets are located nearby.

**Food**
A café situated near the beach serves a variety of food and refreshments.

**Seaside Activities**
Swimming and walking.

**Wet Weather Alternatives**
Scarborough and Filey offer a variety of wet weather activities.

**Wildlife and Walks**
The Cleveland Way runs past the beach, offering several clifftop walks.

**Tourist Information**
Filey Tourist Information Centre, John Street, Filey. Tel. 01723 512204

**Track Record GPGG**
EU designated beach.

*The East Coast*

# TYNEMOUTH LONG SANDS SOUTH
## Tyne and Wear
*OS Ref: NZ369702*

This wide stretch of fine golden sand backed by dunes illustrates all the characteristics of a rural beach in a very urban environment. Its sheer size makes it popular with those wanting to bathe, surf, play beach sports or simply relax. A real jewel, the beach is is greatly enhanced by the adjacent park, sealife centre, shops and man-made rockpool environment.

### Water Quality
No routine sewage discharge has been identified.

### Bathing Safety
Lifeguards patrol the beach between May and September.

### Litter
The beach is cleaned by hand and machine during the summer season and bins are located at strategic points. Dogs are banned from May to September and a poop-scoop scheme operates from October to April.

### Access
Easily accessible from the A19 by car. Access to the beach is via ramps at either end and steps in the centre.

### Parking
There is a pay-and-display car park on the south ramp; further parking is available along the main road behind the beach.

### Public Transport
Frequent bus and Tyne and Wear Metro services travel to and from the city.

### Toilets
These include baby changing and facilities for disabled visitors.

### Food
A wide range of refreshments can be obtained from a variety of nearby cafés, restaurants and hotels.

### Seaside Activities
Swimming, surfing, beach volleyball, canoeing, bowls, boating lake, tennis, deck chair hire and an open-air swimming pool.

### Wet Weather Alternatives
Sealife Centre, amusements and Childhood Memories Toy Museum. Many other venues can be reached by private or public transport.

### Wildlife and Walks
There are coastline walks and extensive rockpool formations at either end of the beach.

### Tourist Information
Whitley Bay Tourist Information Centre, Park Road, Whitley Bay, NE26 1EJ. Tel. 0191 2008535

### Track Record PPGG
EU designated beach.

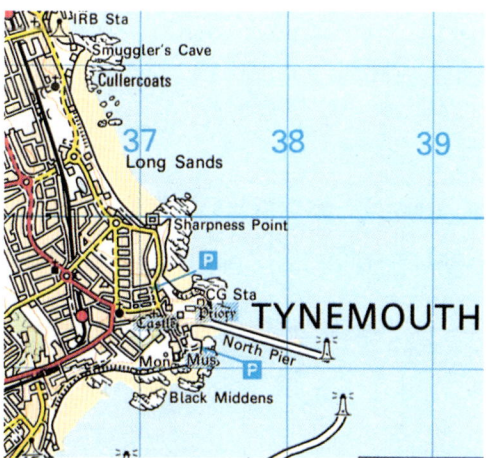

# WARKWORTH
## Northumberland
*OS Ref: NU259065*

Warkworth beach forms part of an extensive expanse of sand that stretches, without interruption, from the mouth of the River Coquet northwards to Almouth Bay. It is backed by an impressive system of dunes, typical of this part of the Northumbrian coastline, which is home to a wide variety of wildlife. The unspoilt qualities of the area are a must for those who enjoy nature, as there is a range of conservation projects and reserves situated along this coast.

### Water Quality
No routine sewage discharge has been identified.

### Bathing Safety
There are no lifeguards and swimmers should be aware of undercurrents. Lifebelts are located at strategic points along the beach.

### Litter
Marine litter is occasionally brought to shore after storms. When badly affected, the beach is cleaned by the local authority.

### Access
The beach is signposted off the A1068. A sandy walkway leads directly from the car park to the beach.

### Parking
There is a car park adjacent to the beach.

### Public Transport
A bus service from Alnwick and Amble runs to Warkworth village, a mile from the beach.

### Toilets
There are toilets located in the beach car park.

### Food
A variety of cafés and pubs in Warkworth village serve food.

### Seaside Activities
Swimming, windsurfing, sailing and fishing.

*The East Coast*

### Wet Weather Alternatives
Warkworth Castle, Alnwick Castle and a number of art galleries are all nearby, as is Barter Books, one of the largest second-hand book shops in England.

### Wildlife and Walks
There are a number of coastal walks in the area. The beach itself forms part of the North Northumberland Heritage Coast and the dunes possess a wide variety of wildlife. There is also a nature reserve at Low Hauxley and boat trips run from Amble around the RSPB reserve at Coquet Island during the summer months.

*Warkworth beach is backed by dunes which exend to the edge of the harbour.*

### Tourist Information
*Seasonal:* Amble Tourist Information Centre, Queen Street, NE65 0DQ. Tel. 01665 712313

*All Year Round:* Alnwick Tourist Information Centre, The Shambles, NE66 1TN. Tel. 01665 510665

**Track Record GGGG**
EU designated beach.

# LOW NEWTON (NEWTON HAVEN)
## Northumberland
*OS Ref: NU242245*

This dune-fringed crescent of sand lies at the northern end of Embleton Bay, overlooked by the pretty village of Low Newton, a square of fishermen's cottages and a pub now owned by the National Trust. Sheltered by a grassy headland to the north and an offshore reef, Low Newton's beach is favoured by watersports fanatics. Newton Pool, a freshwater lagoon behind the dunes, is an important nature reserve.

### Water Quality
No routine sewage discharge has been identified.

### Bathing Safety
Safe on the incoming tide, but swimmers beware of undercurrents on the ebb tide.

### Litter
Occasionally affected by oil drums and fishing debris, more so in winter.

### Access
From the car park on the approach road to Low Newton, signposted off the B1339 from High Newton. It is a short walk down to the village with direct access to the beach. A path leads along Low Newton beach to Embleton Bay.

### Parking
Space for 100 cars between the village and the beach. Village parking is for residents and disabled badge holders only.

*The East Coast*

**Public Transport**
Northumbria Buses run a service to Embleton from Alnwick.

**Toilets**
Toilets are adjacent to the beach.

**Food**
The village pub serves snacks; there is a tea-room in High Newton, ten minutes' walk away.

**Seaside Activities**
Swimming, windsurfing, sailing, diving, canoeing and fishing.

**Wildlife and Walks**
There are bird hides at Newton Pool (one with access for disabled visitors) where a wide variety of species can be seen, particularly in winter. The Heritage Coast

*The curved sweep of Low Newton beach is overlooked by Dunstanburgh Castle, once home to John of Gaunt.*

Path stretches south round Embleton Bay to Dunstanburgh Castle and north round Newton Point to the wide sweep of Newton Links and Beadnell Bay.

**Tourist Information**
*Seasonal:* Craster Tourist Information Centre, The Car Park, Craster, NE66 3TW. Tel. 01665 576007

*All Year Round:* Alnwick Tourist Information Centre, The Shambles, Alnwick, NE66 1TN. Tel. 01665 510665

**Track Record GGGG**
EU designated beach.

## The East Coast

| RATING | NAME | TRACK RECORD | SEWAGE OUTLET | REMARKS |
|---|---|---|---|---|
| | **LINCOLNSHIRE** | | | |
| P | **Skegness** - EU TF572634 | PGPP | 🟩 | 🟨 🗑 Safe bathing. |
| P | **Ingoldmells** - EU TF574685 | PPPP | 1, screened, 120,000, LSO. | 🟨 ⬆ 2000 |
| P | **Chapel St Leonards** - EU TF564722 | PPGP | 🟩 | 🟨 🗑 |
| P | **Anderby Beach** - EU TF553762 | PPPP | 🟩 | 🟨 |
| P | **Moggs Eye** (Huttoft) - EU TF550776 | PGGP | 🟩 | 🟨 🟧 Backed by dunes. |
| G | **Sutton-on-Sea** - EU TF522821 | PGPG | 🟩 | FEATURED |
| P | **Mablethorpe** - EU TF508854 | PPPP | Secondary and UV, 38,000, to Wold Drift Drain. | 🟨 🗑 |
| | **HUMBERSIDE** | | | |
| F | **Cleethorpes** - EU TA310086 | FFFF | Secondary and UV, 70,000, to Tetney Haven. | 🟨 🗑 |
| P | **Withernsea** - EU TA344281 | PPGP | Primary, 7,177, LSO, also stormwater. | 🟨 🟧 |
| P | **Tunstall** - EU TA322312 | PGPP | 🟩 | 🟨 🟧 Backed by low clay cliffs. |
| P | **Hornsea** - EU TA210478 | PPPP | Macerated/screened, 8,050, LSO, also stormwater. | ⬆ 2005 🟨 🟧 |
| P | **Skipsea Sands** - EU TA177572 | PPPP | 🟩 | 🟨 |
| P | **Barmston** - EU TA172594 | PGPP | 🟩 | 🟨 🟧 🗑 |
| P | **Earls Dyke** - EU TA170615 | PFPP | 🟩 | 🟨 🗑 Tidal. |
| P | **Fraisthorpe** - EU TA171629 | GPPP | 🟩 | 🟨 🗑 Tidal. |
| P | **Willsthorpe** - EU TA172640 | PPGP | 🟩 | 🟨 🗑 |
| | **Bridlington** | | | |
| P | South - EU TA181661 | PPGP | Screened, 40,612, LSO, also stormwater. | ⬆ 2000 🟨 🗑 |
| P | North - EU TA190672 | PPGP | Stormwater. | 🟨 🗑 Not featured due to adjacent discharge. Severely affected by industrial discharge |
| | **Flamborough** | | | |
| P | South Landing - EU TA231692 | PPGP | Primary, 3,148, LSO. | 🗑 |
| P | North Landing - EU TA238722 | PPPP | 🟩 | 🟨 🗑 |
| ~ | **Thornwick Bay** | ~~~~ | Primary, 300, at LWM. | ⚠ Bathing very dangerous. Sewage discharge is from a private source. |

🟨 Sand  🟧 Shingle  Pebbles  ⚠ Rocks  🟫 Mud  ❓ No information supplied

# The East Coast

| RATING | NAME | TRACK RECORD | SEWAGE OUTLET | REMARKS |
|---|---|---|---|---|
| | **YORKSHIRE** | | | |
| G | **Reighton Sands** - EU<br>TA144763 | GPGG | 🟩 | FEATURED |
| P | **Filey** - EU<br>TA120806 | PPGP | Raw, 7,138, below LWM, also stormwater. | ⬆ 2000 🟨 🗑 |
| P | **Cayton Bay** - EU<br>TA067845 | FPPP | 🟩 | 🟨 Beware of incoming tide. |
| F | **Scarborough**<br>   South Beach - EU<br>TA046886 | PFGF | 🟩 | 🟨 🗑 Safe bathing. |
| P |    North Beach - EU<br>TA037900 | PPGP | Screened, 52,342, LSO. | ⬆ 2000 🟨 |
| P | **Robin Hood's Bay** - EU<br>NZ959045 | PPFP | Raw, 1,379, at LWM, also stormwater. | ⬆ 2005 ⛰<br>Swimming dangerous. |
| P | **Whitby** - EU<br>NZ897117 | GPGP | Raw, 33,552, at LWM, east of harbour. | ⬆ 2000 🟨 |
| G | **Sandsend** - EU<br>NZ864126 | PPPG | Raw, 511, at LWM, also stormwater. | ⬆ 2005 🟨 🟧 Not featured due to adjacent discharge. |
| P | **Runswick Bay** - EU<br>NZ811159 | PPPP | Raw, 591, at LWM. | ⬆ 2005 🟨 🟧 |
| F | **Staithes** - EU<br>NZ787190 | PPPF | Raw, 1,718, at LWM, also stormwater. | ⬆ 2005 🟨 ⛰ |
| | **CLEVELAND** | | | |
| ~ | **Skinningrove**<br>(Cattersty Sands)<br>NZ710205 | FP~~ | Raw, 9,000, at LWM. | 🟨 |
| P | **Saltburn-by-the-Sea** (Pier) - EU<br>NZ666217 | FPPP | 🟩 | 🟧 |
| P | **Marske-by-the-Sea** - EU<br>NZ636232 | PPPP | 2, stormwater only, 25m below and at LWM. | 🟨 🗑 |
| P | **Redcar**<br>   Stray - EU<br>NZ625238 | PPPP | Screened, 78,000, LSO. | ⬆ 🟨 |
| F |    Granville - EU<br>NZ613251 | PPPF | Screened, stormwater, at LWM. | 🟨 |
| P |    Lifeboat Station - EU<br>NZ606255 | PPPP | Screened, stormwater, at LWM. | 🟨 ⛰ |
| P |    Coatham Sands - EU<br>NZ592257 | PPPP | 2, screened, both stormwater, at LWM. | 🟨 🗑 |
| P | **Seaton Carew**<br>   North Gare - EU<br>NZ540286 | PPPP | (see Seaton Carew - Centre) | 🟨 Not featured due to adjacent discharge. |
| P |    Centre - EU<br>NZ531296 | FPGP | Screened, 60,000, 300m below LWM and LSO. Stormwater during very high flows only. | 🟨 🗑 |
| P |    North - EU<br>NZ525305 | PPPP | (see Seaton Carew - Centre) | 🟨 |
| ~ | **Hartlepool**, North Sands<br>NZ513350 | ~FP~ | 2 raw, 1 screened, 7,900 (raw), 30,000 (screened), 30m below LWM and 2 at LWM. | 🟨 |

🟩 No discharge identified   ⬆ Improvements planned   ❓ Insufficient information to feature   🗑 Cleaned regularly

## The East Coast

| RATING | NAME | RECORD | TRACK OUTLET | SEWAGE REMARKS |
|---|---|---|---|---|
| | **COUNTY DURHAM** | | | |
| F | **Crimdon Park** - EU<br>NZ485373 | FFFF | Improvements ongoing. | ⬆️ 🟨 |
| ~ | **Crimdon South**<br>NZ495363 | F~~~ | Stormwater only. | 🟨 |
| ~ | **Blackhall**<br>NZ472393 | ~~~~ | Stormwater only, at LWM. | 🟨 Pebbles |
| | **Denemouth South**<br>NZ457407 | ~~~~ | Screened, 33,000, at LWM. | ⬆️ 🟨 Stones and coal waste. |
| | **Horden**<br>NZ447425 | ~~~~ | 2, raw, 18,400 (total), both above LWM. | 🟨 Coal waste. |
| ~ | **Easington**<br>NZ445440 | P~~~ | Raw, 8,300, 50m below LWM. | ⬆️ 🟨 Stones and coal waste. |
| ~ | **Dalton Burn**<br>NZ445440 | F~~~ | ❓ | ⬆️ Coastal stream. |
| | **Seaham** | | | |
| F | Remand Home - EU<br>NZ426505 | FFPF | Stormwater only. | ⬆️ 🟨 |
| F | Beach - EU<br>NZ424508 | FFPF | ❓ | ⬆️ 2000 🟨 |
| ~ | **Featherbed Rocks** | F~~~ | 🟩 | 🔺 Adjacent to mouth of Dalton Beck. |
| | **TYNE AND WEAR** | | | |
| | **City of Sunderland** | | | |
| ~ | Ryhope South<br>NZ417530 | F~~~ | 2, stormwater only, both at LWM. | 🟨 |
| ~ | Hendon South<br>NZ412555 | F~~~ | ❓ | |
| ~ | Sunderland | ~~~~ | 1 raw, 1 screened, 175,000, 500, at LWM, 300m below LWM. | Rocky outfalls situated south of Wear Estuary. |
| P | **Roker/Whitburn** South – EU<br>NZ407593 | PPPP | ❓ | 🟨 Pebbles |
| ~ | **Roker/Blockhouse** | F~~~ | Screened, 25m below LWM. Discharge within Wear Estuary. | 🟨 |
| G | **Whitburn** North - EU<br>NZ407605 | FFPG | Stormwater only, at LWM. | 🟨 |
| F | **Marsden Bay** - EU<br>NZ400650 | FPPF | Raw from public toilet, also stormwater | 🟨 Not featured due to adjacent discharge |
| P | **South Shields** (Sandhaven) - EU<br>NZ379674 | FPPP | 🟩 | Beach backed by sand dunes and a small cliff. |
| | **Tynemouth** | | | |
| G | King Edward's Bay - EU<br>NZ373696 | PPGG | Stormwater only, at LWM. | 🟨 Not featured due to adjacent discharge. |
| G | Long Sands South - EU<br>NZ369702 | PPGG | 🟩 | FEATURED |
| P | Long Sands North - EU<br>NZ366708 | PPGP | Stormwater only, at LWM. | 🟨 |
| P | Cullercoats - EU<br>NZ365713 | PPPP | 🟩 | 🟨 🗑️ |
| P | **Whitley Bay** - EU<br>NZ353734 | PFGP | 10, stormwater, all at LWM. | 🟨 🔺 |

🟨 Sand  Shingle  Pebbles  Rocks  Mud ❓ No information supplied

*The East Coast*

| RATING | NAME | TRACK RECORD | SEWAGE OUTLET | REMARKS |
|---|---|---|---|---|
| G | **Seaton Sluice** - EU<br>NZ334771 | FPPG | Stormwater only. | 🟨 🗑 Not featured due to adjacent discharge. |
| | NORTHUMBERLAND | | | |
| P | **Blyth**<br>  South Beach - EU<br>  NZ322795 | FPPP | 🟩 | 🟨 🗑 Dunes. |
| ~ | **Cambois**<br>  South<br>  NZ310835 | F~~~ | Raw, 26,000, 50m below LWM. | 🟨 🟧 |
| ~ | North | F~~~ | (see Cambois South) | 🟨 🟧 |
| G | **Newbiggin**<br>  South - EU<br>  NZ311873 | GPGG | (see New Biggin - North) | 🟨 Not featured due to adjacent discharge. |
| P | North - EU<br>NZ313878 | PPPP | Emergency outfall. | 🟨 |
| ~ | **Cresswell**<br>NZ290940 | ~~~~ | 1,800 (summer), 200 (winter), at LWM. | 🟨 |
| P | **Druridge Bay** - EU<br>NZ279964 | PGGP | 🟩 | 🟨 |
| G | **Amble** (Links) - EU<br>NU276044 | PGGG | Macerated/screened, 8,000, 250m below LWM. | 🟨 ⛰ Not featured due to adjacent discharge. |
| G | **Warkworth** - EU<br>NU259065 | GGGG | 🟩 | FEATURED |
| P | **Alnmouth** - EU<br>NU253107 | PGGP | 🟩 | 🟨 |
| ~ | **Longhoughton Steel**<br>NU263157 | ~~~~ | Raw, 1,200 (summer), 200 (winter), at LWM. | 🟨 |
| ~ | **Craster**<br>NU260200 | ~~~~ | 2, raw and stormwater, 400, 30m below LWM. | |
| ~ | **Embleton Bay**<br>NU245225 | ~~~~ | 🟩 | 🟨 |
| G | **Low Newton**<br>(Newton Haven) - EU<br>NU242245 | GGGG | 🟩 | FEATURED |
| P | **Beadnell Bay** - EU<br>NU233284 | PGPP | 🟩 | 🟨 |
| P | **Seahouses** - EU<br>NU211330 | GGGP | 🟩 | 🟨 ⛰ Dunes, bathing safe on incoming tides only. |
| G | **Bamburgh** - EU<br>NU185353 | GGGG | Macerated, 1,000 (summer), 700 (winter), at LWM. | 🟨 Not featured due to adjacent discharge. |
| ~ | **Holy Island** (Lindisfarne) | ~~~~ | Macerated, 500 (summer), 200 (winter), at LWM. | 🗑 |
| ~ | **Cocklawburn Beach**<br>NU030485 | ~~~~ | Raw, 15, 20m above LWM. | 🟨 ⛰ |
| P | **Spittal** - EU<br>NU008515 | PPPP | Secondary. | 🟨 Beware of ebbing tide. |
| ~ | **Spittal Quay** | F~~~ | (see Spittal) | 🟨 |
| ~ | **Berwick-upon-Tweed** | F~~~ | Secondary. | 🟨 Swim only between rocks and beach. |

🟩 No discharge identified   ⬆ Improvements planned   📛 Insufficient information to feature   🗑 Cleaned regularly

# THE NORTH-WEST

FROM SKINBURNESS IN CUMBRIA DOWN TO WEST KIRBY ON THE WIRRAL AND INCLUDING THE ISLE OF MAN, THE NORTH-WEST COAST OF ENGLAND IS THE REGION THAT PIONEERED THE SEASIDE RESORT AND THE SEASIDE HOLIDAY. IT SHOULD THEREFORE BE A SOURCE OF NATIONAL SHAME THAT NO BEACHES IN THIS AREA ARE GOOD ENOUGH TO BE FEATURED IN THE *GOOD BEACH GUIDE*.

•

It is not that there aren't many beautiful stretches of coastline along the North-West. The huge expanse of Morecambe Bay, for example, with its panoramic view over to the Lake District; Southport, its long beach and elegant buildings lending it a distinctly Victorian character; and the extensive dune system of the Ainsdale National Nature Reserve, a habitat for rare natter-jack toads and sand lizards, which has over 10 kilometres of clearly marked footpaths, dunes and pine woods down to National Trust land at Formby Point.

In the middle of the Irish Sea lies the Isle of Man, a fascinating destination often described as an island lost in time. It has an incredible variety of coastal scenery for such a small island, its fiercely rugged coastline contrasting with the delicate beauty of the Manx Glens. In the summer huge but gentle basking sharks are to be seen offshore, feeding on plankton.

But the sad truth is that the entire region is still badly affected by pollution, indelibly tainting its underlying beauty. The Irish Sea is now more chemically contaminated than the North Sea and, perhaps more insidiously, is affected by radioactive waste. The Mersey Estuary and Liverpool Bay have suffered particularly, not only from oil spills and incidents of accidental pollution, but from the deliberate and unacceptable discharges of mercury, cadmium and lead released into Liverpool Bay each day. This contaminated sea washes the shore of some the most popular resorts in Britain, and as a result their beaches are amongst the most polluted in the whole of the country. Unhappily, visitors must expect to witness this pollution in some if its most distressing forms: sewage-related debris on beaches and illegally dumped refuse washed up on the shore.

Despite ongoing investment in sewage treatment, the legacy of under-investment means that only slight improvement has been apparent in the North-West over the past year, and this after the quality of bathing water at a number of the region's beaches actually deteriorated during 1995. North West Water, in a move welcomed by the Marine Conservation Society, has nearly completed work on a system for Blackpool that will provide full sewage treatment, but bathing is still not advisable at most beaches in this area. Certain beaches at Blackpool have met minimum legal standards this year for the first time, but they do not reach Guideline standards and cannot therefore be recommended. It appears that the problems in the North-West are far from over.

The Isle of Man fares little better: although not as badly affected by industrial discharges as the mainland coast, they do have a problem with sewage. The island's sewage is discharged directly to sea via short sea outfalls, most with absolutely no treatment: unsurprisingly the beaches are badly contaminated. The EC Bathing Water Directive does not apply to the Isle of Man, but in March 1990 the Manx Government decided that the EC standards should be adopted as a target. In 1996, however, only four bathing water passed the minimum requirements when tested.

*The Calf of Man is particularly known for the richness and diversity of its marine life.*

This situation should change dramatically with the coming on stream of the IRIS project, an ambitious scheme that aims to achieve Integrated Recycling of the Island's Sewage. The scheme should make the Isle of Man's beaches some of the cleanest in the UK and perfectly illustrates how a lamentable situation can be radically improved with a little imagination and the will to change. We are sorry to report, however, that progress on the scheme has been slow and it is still some way from completion.

## The North-West

| RATING | NAME | TRACK RECORD | SEWAGE OUTLET | | REMARKS |
|---|---|---|---|---|---|
| | **CUMBRIA** | | | | |
| F | **Skinburness (Silloth)** - EU NY126565 | PPPF | 🟩 | | 🟨 |
| F | **Silloth (Lees Scar)** EU NYO94528 | PPPF | | Secondary, UV, 7,000, 50m below LWM. | 🟨 🟧 |
| P | **Allonby South** EU NY066406 | FFPP | 🟩 | | 🟨 ⛰ 🟫 |
| P | **Allonby South - West Winds** EU NY078424 | PPPP | | Secondary 1,100, 270m below HWM. Storage for ebb tide discharge. | 🟨 |
| ~ | **Maryport** NY028356 | ~~~~ | 🟩 | | 🟨 🟧 |
| ~ | **Flimby** NY018334 | ~~~~ | 🟩 | | 🟨 🟧 |
| ~ | **Siddick** NY001316 | ~~~~ | | Primary 17,000, 2500m below LWM. | |
| ~ | **Workington** NX983296 | ~~~~ | | 4 raw, 26,000 (total) at LWM, 10 and 20m above, below LWM. | ⬆ 2000. Shingle and slag. Low amenity. |
| ~ | **Harrington** | ~~~~ | 🟩 | | Shingle and slag. |
| ~ | **Parton** | ~~~~ | | Screened, 200 + industry, 800m below LWM. | ⬆ 2000 🟧 Low amenity. |
| ~ | **Whitehaven** | ~~~~ | | Screened, 27,000, 150m below HWM. | ⬆ 2000 🟧 Little used. |
| P | **St Bees** - EU NX959117 | PPFP | | Primary 2,000, 250m below HWM. | 🟨 🟧 |
| ~ | **Nethertown** | ~~~~ | | Raw, 500, at LWM. | ⬆ 2005 🟧 |
| ~ | **Braystones** | ~~~~ | | Raw, 9,000, 50m below LWM. | ⬆ 2005 🟧 |
| P | **Seascale** - EU NY034010 | FFFP | | Primary, 2,200, 1,500m below LWM. | 🟨 🟧 ⛰ |
| ~ | **Ravenglass** | ~~~~ | | Primary, 250+ tourists, to Esk at LWM. | 🟧 🟫 |
| P | **Silecroft** - EU SD120812 | PPPP | 🟩 | | 🟨 🟧 |
| F | **Haverigg** - EU SD157766 | FPFF | 🟩 | | 🟨 Dunes, high amenity. |
| ~ | **Millom** | ~~~~ | | Secondary, 7,500 at LWM. | 🟨 🟧 |
| F | **Askam-in-Furness** - EU SD210788 | FPFF | | Secondary, UV, 2,350, to stream at HWM. | 🟨 Bathing safe inshore. |
| ~ | **Barrow-in-Furness** | ~~~~ | | Many raw, primary and secondary, 720,000 (total) to Walney Channel. | 🟨 |
| P | **Roan Head** - EU SD198758 | PFFP | 🟩 | | 🟨 |
| F | **Walney Island - West Shore** EU SD170700 | PFFF | 🟩 | | Beaches to the west of the island have sand dunes and normal bathing facilities. |

🟨 Sand  🟧 Shingle  🟦 Pebbles  ⛰ Rocks  🟫 Mud  ❓ No information supplied

# The North-West

| RATING | NAME | TRACK RECORD | SEWAGE OUTLET | REMARKS |
|---|---|---|---|---|
| P | **Walney Island - Sandy Gap** EU SD175681 | PGPP | 🟩 | 🟨 |
| P | **Walney Island - Biggar Bank** EU SD178673 | PPPP | 🟩 | 🟨 |
| P | **Newbiggin** - EU SD273694 | FPFP | Secondary, UV, to stream at HWM. | 🟨 |
| F | **Aldingham** - EU SD283709 | FPPF | 🟩 | 🟨 |
| F | **Bardsea** - EU SD300740 | FPFF | 🟩 | 🟨 Country park. |
| ~ | **Grange-over-sands & Kents Bay** | ~~~~ | Secondary, 11,500, to Wyke Bay. | 🟨 🟫 🟤 |
| ~ | **Arnside** | ~~~~ | 2,000, tidal tank. | 🟨 🟫 🟤 |
| | LANCASHIRE | | | |
| ~ | **Hest Bank** | ~~~~ | 2, secondary, 2,850 (total), above and below HWM. | 🟤 Sea retreats to 6.5km, no bathing. |
| F | **Morecombe - North** - EU SD441650 | FFFF | (see Morecombe South) | 🟫 🟤 |
| P | **Morecombe - South** - EU SD422636 | FPFP | Screened, 31,000, at LWM. | ⬆ 1997 🟨 |
| F | **Heysham - Half Moon Bay** - EU SD413618 | PPFF | 🟩 | 🟨 Popular beach. |
| ~ | **Pilling Sands** | ~~~~ | Primary 1,000, to stream. | 🟤 Salt marsh. |
| ~ | **Knot End-on-Sea** | ~~~~ | 🟩 | 🟨 🟤 |
| F | **Fleetwood (Pier)** - EU SD336458 | FFPF | Secondary, 330,000 and stormwater, 3800m below LWM. | 🟨 |
| P | **Cleveleys** - EU SD312433 | FPFP | 🟩 | 🟨 |
| P | **Bispham** - EU SD307397 | FPPP | 🟩 | 🟨 |
| P | **Blackpool - North Pier** - EU SD305364 | FPFP | 🟩 | ⬆ 🟨 |
| P | **Blackpool - Central (Lost Children's Post)** - EU | FFFP | 🟩 | ⬆ 🟨 |
| F | **Blackpool - South Pier** - EU SD304305 | FFFF | 🟩 | ⬆ 🟨 |
| F | **St Anne's - North** EU SD304305 | FFFF | 🟩 | 🟨 |
| F | **Lytham St Anne's** - EU SD318283 | FFFF | 🟩 | 🟨 🗑 |
| | MERSEYSIDE | | | |
| P | **Southport** EU SD322179 | FPFP | Secondary, 115,000, to Crossens Pool. | 🟨 Lifeguard and first aid service in season. |
| P | **Ainsdale** EU SD297129 | FPPP | 🟩 | 🟨 Lifeguard and first aid service in season. |

🟩 No discharge identified  ⬆ Improvements planned  ⁇ Insufficient information to feature  🗑 Cleaned regularly

## The North-West

| RATING | NAME | TRACK RECORD | SEWAGE OUTLET | REMARKS |
|---|---|---|---|---|
| P | **Formby** - EU SD277100 | PPPP | ■ | ■ |
| ~ | **Hightown** | ~~~~ | Stormwater. | ■ ■ |
| ~ | **Blundell Sands** | ~~~~ | 2, 48,100 (total) | ■ |
| P | **New Brighton (Harrison Drive)** EU SJ287937 | FPPP | ■ | ■ ■ |
| P | **Moreton** - EU SJ257918 | GGPP | Screened, 65,000, LSO. | ⬆ 2000 |
| P | **Meols** - EU SJ230906 | PPPP | (as Moreton) | ■ |
| F | **West Kirby** - EU SJ210868 | FPFF | Primary, 10,500, LSO, secondary treatment during 1997. | ■ |
| | ISLE OF MAN | | | |
| P | **Douglas - Summerhill** | F~PP | 3, raw, 290,000 (total), 600m below and at LWM also stormwater. | ⬆ ■ ■ ▲ Beach running northwards from river mouth, islands. |
| F | **Douglas - Palace** | F~FF | (see Douglas - Summerhill) | (see Douglas Summerhill) |
| F | **Douglas - Broadway** | F~FF | (see Douglas - Summerhill) | (see Douglas Summerhill) |
| F | **Laxey** | F~PF | Raw, 1,500, below LWM. | ⬆ ■ below half tide. ■ |
| F | **Ramsey** | F~FF | Raw, 6,500, below LWM. | ⬆ ■ Two beaches either side of old working harbour. |
| P | **Peel** | F~FP | Raw, 3,800 50m below LWM. | ⬆ ■ Old working harbour and Peel Castle. |
| F | **White Strand** | ~~~F | ■ Affected by Peel outfall. | |
| P | **Fenella Beach** | ~~~P | ■ Affected by Peel outfall. | |
| F | **Port Erin** | F~~F | Raw, 3,000, below LWM. | ⬆ ■ Beach in sheltered bay on south of island. |
| F | **Port St Mary** | F~FF | 2, raw, 2,000 (total), both below LWM. | ⬆ ■ |
| F | **Castletown** | F~FF | 2, raw, 3,200 (total), both below LWM. | ⬆ ■ ■ ▲ Used extensively for watersports. |
| F | **Derbyhaven** | P~PF | Raw, 150, below LWM (pumped tidal discharge). | ⬆ ■ ■ Shingle area used for mooring. |
| F | **Gansey Bay (Bay Ny Carrickey)** | ~~FF | ■ | ■ ■ Watersports centre. |
| F | **Kirk Michael** | F~FF | Screened, primary, 420, below LWM. | ⬆ ■ Not extensively used for bathing. |
| F | **Jurby** | ~~PF | Primary, 480, below LWM. | ⬆ Bathing not advised due to currents. |
| P | **Port Soderick** | ~~~P | No public discharges, private septic tanks. | ■ |

■ Sand  ■ Shingle  ■ Pebbles  ▲ Rocks  ■ Mud  ? No information supplied
■ No discharge identified  ⬆ Improvements planned  ?? Insufficient information to feature  ⌑ Cleaned regularly

# SCOTLAND

SCOTLAND HAS SOME OF THE MOST SPECTACULAR COASTAL SCENERY IN THE WORLD, FROM THE LONG SAND DUNES OF THE EAST COAST TO THE ROCKY SHORES OF FIFE WITH ITS SERIES OF PICTURESQUE FISHING VILLAGES. THIS CHAPTER COVERS THE COAST OF MAINLAND SCOTLAND: BEACHES THAT ARE RELATIVELY EASY TO REACH FOR A DAY AT THE SEA, OR AS A STARTING POINT TO EXPLORE FURTHER THE DELIGHTS OF THIS COASTLINE. INFORMATION ON THE HEBRIDES, ORKNEY AND SHETLAND HAS NOT BEEN INCLUDED AS BATHING WATER QUALITY DATA FOR THESE AREAS IS NOT AVAILABLE. BEACHES ABOUND ON THE OUTER ISLANDS, HOWEVER, AND MANY OF THEM ARE PRISTINE WITH FEW OR NO WATER QUALITY PROBLEMS.

•

From the cliffs and stacks of Caithness to the west coast Highlands and Islands, with sea lochs and towering mountains in between, there are hundreds (if not thousands) of beaches and tiny sandy bays, mostly remote, deserted and beautiful. Many can only be reached by the keen walker, but the effort usually proves well worth while. Often, the remoteness of Scottish beaches – particularly those of the northern and western coast – means that they are free from the pollution which taints many of Britain's other shores. As they are not routinely monitored for sewage pollution, however, we are unable to recommend them by name in the *Guide*. Nonetheless, if what you seek is peace and solitude combined with traditional hospitality, then Scotland may be the place for you.

That is not to say that Scotland does not have problems around its coast. Many Scottish beaches have failed to meet the minimum EC bathing water quality standard. Sea-borne rubbish is washed up on the shore; sewage sludge and dredged spoil are dumped off the Clyde, Forth and Tay Estuaries; industrial waste is discharged into coastal waters, particularly around the Clyde and the Forth. As in England, nuclear installations contribute to pollution of the sea by discharging warm water and contamination from anti-fouling treatments of the cooling water intake pipes.

The rapid development of the coastline has, seemingly inevitably, destroyed once scenic areas. The sea lochs of the west coast are studded with the telltale floating cages of the fish farming business. The North Sea oil and gas industry has spawned the growth of massive terminals and the view along the Cromarty Firth is dominated by a string of platforms. The recent discovery of a large oil-field off the west coast of Scotland is a cause for concern since its exploitation will lead to increased shipping and increased risk of accidents in an area of outstanding natural beauty, and of international importance for birds, seals and fish. Any future development of this oil-field must be carefully controlled and managed to ensure that the risk of accidents and routine pollution is kept to a minimum. The Marine Conservation Society is campaigning hard to ensure that adequate provisions for the protection of the environment remain part of the ongoing development of the UK's offshore oil and gas industry.

# PEASE SANDS
## Borders
*OS Ref: NT794710*

A cove of red cliffs with a large expanse of tawny sand is backed by a caravan park, beyond which lie the glorious wooded glens of Pease Dean and Tower Dean.

### Water Quality
One outfall discharges secondary level sewage from the local caravan park above LWM.

### Bathing Safety
Generally safe; no lifeguard patrol.

### Access
Follow the signs to Pease Sands from the A1 at Cockburns Path. Access to the beach is via the caravan park.

### Parking
At the caravan park.

### Toilets
There are toilets in the caravan park.

### Food
From the caravan park.

### Seaside Activities
Swimming.

### Wildlife and Walks
Pease Dean wildlife reserve.

### Tourist Information
Eyemouth Tourist Information Centre, Market Square, Eyemouth. Tel 018907 50678 (summer only).

**Track Record PPPG**
EU designated beach.

*Torness nuclear power station makes an unusual backdrop for the beach at Thorntonloch.*

*Scotland*

# THORNTONLOCH
## Lothian
*OS Ref: NT753746*

Almost hidden behind the rocky outcrop of Torness Point and within sight of the nuclear power station, this beautiful sandy beach runs south for about 400m, backed along its length by shallow dunes. To the south are spectacular views of the Berwickshire cliffs, and close by the coastline runs a well marked Geology Trail.

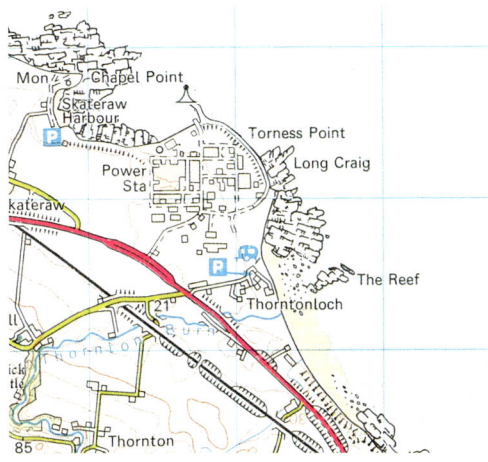

### Water Quality
No routine sewage discharge has been identified.

### Bathing Safety
A strong undertow means swimmers should exercise extreme caution. There are no lifeguards on duty.

### Litter
Cleaned weekly throughout the summer season.

### Access
The beach is on an unclassified road off the A1, south of the entrance to Torness Nuclear Power Station.

### Parking
A car park close by the caravan site and only a short walk from the beach has space for approximately 40 cars.

### Toilets
There are toilets on the caravan park.

### Food
Food is available on the caravan site adjacent to the car park.

### Seaside Activities
Swimming, windsurfing and fishing.

### Wet Weather Alternatives
Dunbar Tourist Information has details of local attractions.

### Wildlife and Walks
There is a birdwatching site at Barns Ness, 3 km to the north; the Geology Trail starts in the car park at White Sands Bay (p136).

### Tourist Information
Dunbar Tourist Information Centre, Dunbar. Tel. 01368 863353

**Track Record PGGG**
Not EU designated.

# WHITE SANDS BAY
## Lothian
*OS Ref: NT710773*

An attractive, compact beach with fine golden sand bounded by a rocky shore and backed by low coastal grassland spotted with picnic tables, White Sands is an ideal place for an early-morning swim. There are pleasant views to be had along the coast towards the old town of Dunbar in the distance.

**Water Quality**
No routine sewage discharge has been identified.

**Bathing Safety**
Swimming is safe in all but the most extreme weather conditions. There is no lifeguard cover.

**Litter**
The beach is cleaned weekly during the summer season.

**Access**
White Sands is clearly signposted off the A1, approximately five kilometres south of Dunbar. Access to the beach can be difficult for disabled visitors.

**Parking**
Ample space on grassland.

**Public Transport**
The nearest railway station is Dunbar; a bus runs from the town to the cement works, which is only a short walk from the beach.

**Seaside Activities**
Swimming.

**Wet Weather Alternatives**
Dunbar Tourist Information has details of local attractions.

**Wildlife and Walks**
The Geology Trail starts and ends in

*Scotland*

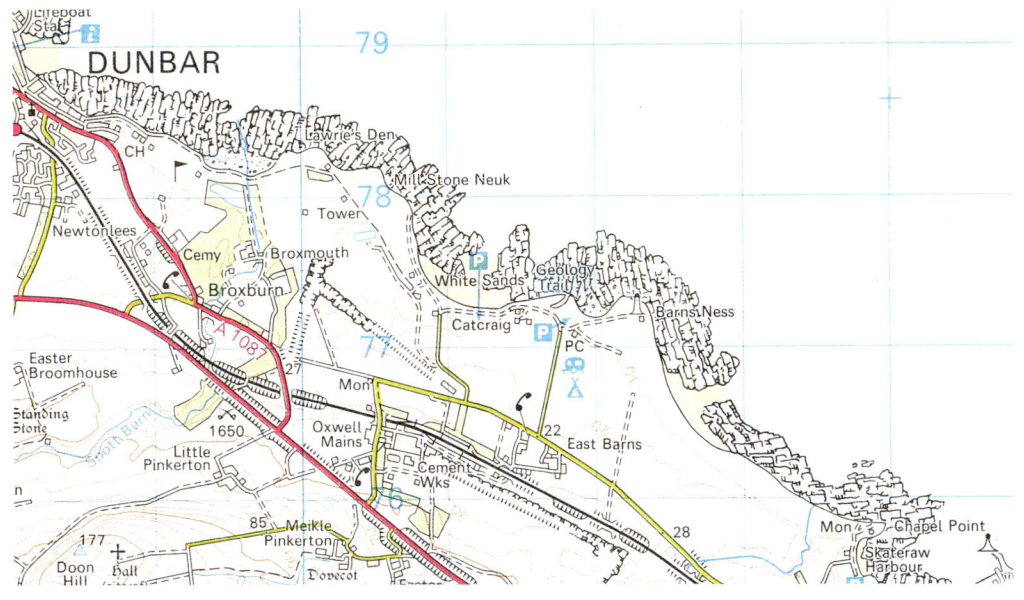

the car park and this is a good base for easy coastal walks and rambles. There is excellent birdwatching: the local pools and bushes attract a wide range of species especially at migration time.

**Tourist Information**
Dunbar Tourist Information Centre, Dunbar. Tel. 01368 863353

**Track Record PGGG**
Not EU designated.

*Wild flowers abound in the grasslands backing the beach at White Sands Bay.*

# ELIE/EARLSFERRY BEACH
## Fife
*OS Ref: NO485998*

Straddling a natural sandy bay that faces south across the Firth of Forth towards Bass Rock lie the linked settlements of Elie and Earlsferry. Once renowned respectively as a fishing port and a market town, Elie and Earlsferry now find popularity amongst those holidaying on the coast of Fife. The beach itself is composed of dark-golden sand backed by dunes, and its sheltered position means that it provides a safe location for bathing. In addition to the local towns, there are also a number of attractions located in the surrounding area, including the Scottish Fisheries Museum at Anstruther.

**Water Quality**
No routine sewage discharge has been identified in the vicinity of this beach.

**Bathing Safety**
Bathing here is considered safe. Lifeguards patrol the beach from June tol September.

**Litter**
The beach is cleaned frequently (on a daily basis during the summer). Dogs are restricted to a specific exercise area between June and September.

**Access**
Elie can be reached on the A917, leading west from Kirkaldy. The beach is accessed from the road and car park.

**Parking**
There is parking in the town in addition to the car park adjacent to the beach.

*Scotland*

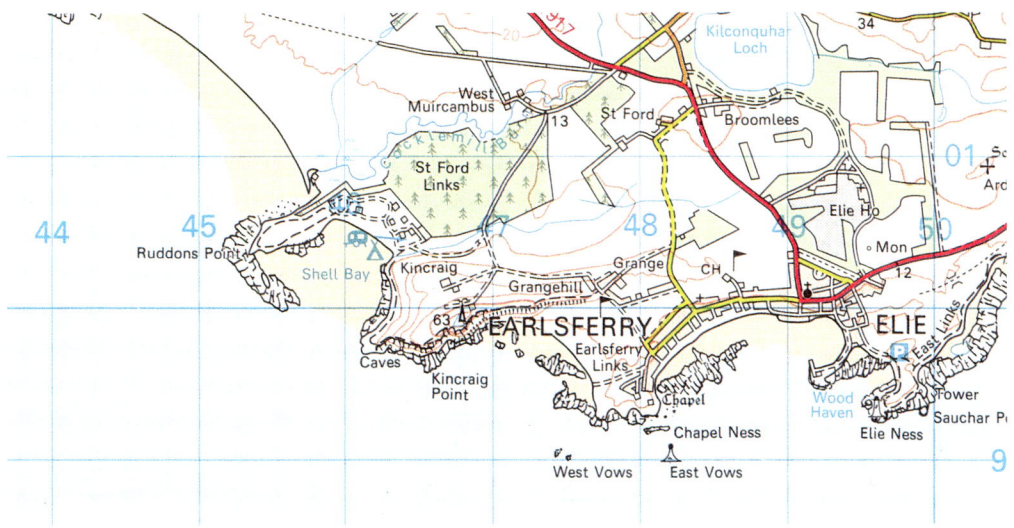

### Public Transport
A bus service from Kirkcaldy to St Andrews calls at Elie.

### Toilets
Toilets are located adjacent to the beach. There are also facilities for the disabled.

### Food
There are a number of cafés and shops situated in Elie.

### Seaside Activities
Swimming is particularly recommended at this beach. In addition a locally based watersports centre offers tuition and hire of equipment for a range of activities.

### Wet Weather Alternatives
Kellie Castle and the Scottish Fisheries Museum, Anstruther. There is also a recreation centre at Anstruther and St Andrews – the spiritual home of golf – is only 19 kilometres away.

### Tourist Information
St Andrews and the Kingdom of Fife Tourist Board, 70 Market Street, St Andrews, Fife, KY16 9NU.
Tel. 01334 477872

**Track Record PPGG**
Not EU designated.

*These sands lie in a sheltered bay bounded by the settlements of Elie and Earlsferry.*

# ROOME BAY (CRAIL)
## Fife
*OS Ref: NO618078*

This small sandy beach backed by a steep grassy park is but a short walk from the pretty and unspoilt harbour town of Crail. It has enough rock pools to satisfy the curiosity of any budding marine biologist.

**Water Quality**
No routine sewage discharge has been identified.

**Bathing Safety**
Bathing is safe in Old Rock Swimming Pool. There is no lifeguard cover.

**Litter**
The beach is cleaned weekly during the summer by machine, and litter bins are positioned along the beach. There is no dog ban.

**Access**
Follow the A917 from St Andrews or take the Coastal Tourist Route. The beach is reached from the town, but disabled visitors might find the going difficult as access can only be gained via steep tarmac paths and steps.

**Parking**
Space at Roome Bay is limited, but there is ample parking in the town centre some 15 minutes walk from the beach.

**Public Transport**
The town is served by a bus service from St Andrews and Edinburgh.

**Toilets**
Toilets are located nearby.

**Food**
Widely available in the town.

**Seaside Activities**
Swimming and rockpooling.

**Wet Weather Alternatives**
Scottish Fisheries Museum, Anstruther. There are also many activities to be enjoyed at St Andrews.

**Wildlife and Walks**
The footpath around the Fife coast can be joined here.

**Tourist Information**
St Andrews and the Kingdom of Fife Tourist Board, 70 Market Street, St Andrews, Fife, KY16 9NU.
Tel. 01334 477872

**Track Record PPGG**
Not EU designated.

# PETERHEAD LIDO
## Grampian
*OS Ref:NK123451*

The quality of the sand and water belie the busy but interesting harbour environment. Peterhead is Europe's largest white-fish port, so it is somewhat surprising to find this jewel of a beach within its protective sea walls. Adjacent to the beach is the award-winning Maritime Heritage Centre and a caravan park, marina and sailing club.

**Water Quality**
No routine sewage discharge has been identified.

**Bathing Safety**
Bathing is generally safe if normal precautions are adhered to; there is no lifeguard cover.

**Litter**
The beach and dunes are cleaned regularly. Dogs are banned from the beach between April and September unless they are on a lead. A poop-scoop scheme is in operation, and bins are provided at the entrance to the beach.

**Access**
The beach is 2 kilometres south of Peterhead town centre on the A952. Steps and disabled access to the beach lead from the adjacent car park.

**Parking**
There is parking for 120 cars and provisions are made for coaches.

**Public Transport**
The Aberdeen to Peterhead buses stop at the beach, along with the town service bus.

**Toilets**
These include facilities for disabled visitors.

**Food**
There is a café at the Maritime Heritage Centre adjacent to the beach.

**Seaside Activities**
Swimming, surfing, windsurfing, sailing, and children's play area.

**Wet Weather Alternatives**
The Maritime Heritage Centre; the Arbuthnot museum; the 18th-century Town House; the Fish House.

**Tourist Information**
Aberdeen and Grampian Tourist Board, North Silver Street, Aberdeen, AB1 1RJ.
Tel. 01224 848848

**Track record PPGG**
Not EU designated.

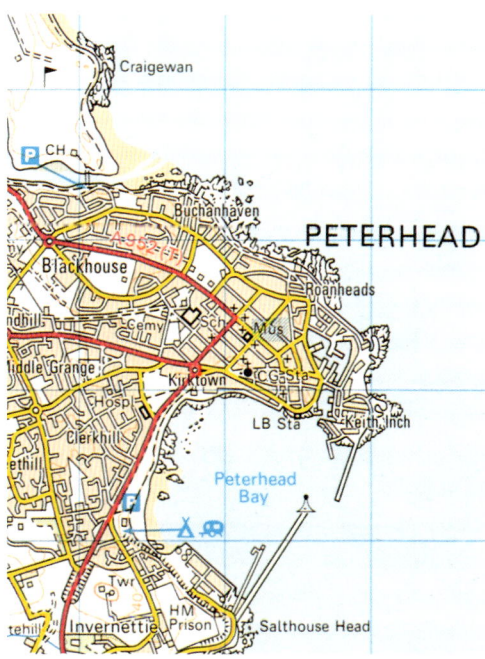

# BALMEDIE
## Grampian
*OS Ref:NJ9818*

Just north of Aberdeen lies one of Britain's longest sandy beaches, extending from the River Don on the city's northern outskirts to the mouth of the River Ythan 16 kilometres to the north. The beach is backed by extensive dunes, and the fine sand shifts continuously. Shell spotters will find an abundance of fine specimens, both common and exotic, at the tideline. Towards the southern end of the beach a rifle range at Blackdog flies red flags when the range is in use. The sheer size of the beach means it is seldom crowded and makes it possible to find a quiet spot at any time of year.

### Water Quality
One outfall in the vicinity discharges secondary treated sewage through a long sea outfall via pumped diffusers.

### Bathing Safety
Swimming is generally safe along the whole of the beach, with the usual precautions.

### Litter
Occasionally severely affected by deposits of marine debris, but cleaned regularly between April and September.

### Access
The main point of access is at Balmedie, signposted from the A92.

### Parking
There are several car parks at Balmedie in the Country Park.

### Seaside Activities
Swimming.

### Wet Weather Alternatives
Old Slains Castle to the north; Crathes Castle and Gardens to the west;

several museums in Aberdeen to the south.

### Wildlife and Walks
The sands of Forvie Nature Reserve at the northern end of Balmedie Beach, which includes the Ythan Estuary, are home to kittiwakes, terns, geese and ducks, and contains the largest colony of eider ducks in Britain. Access is via a car park on the A975 road north of Newburgh. A themed guided walks programme is run by the Aberdeenshire Ranger Service.

### Tourist Information
Aberdeen and Grampian Tourist Board, North Silver Street, Aberdeen, AB1 1RJ.
Tel. 01224 848848

### Track record PPGG
Not EU designated.

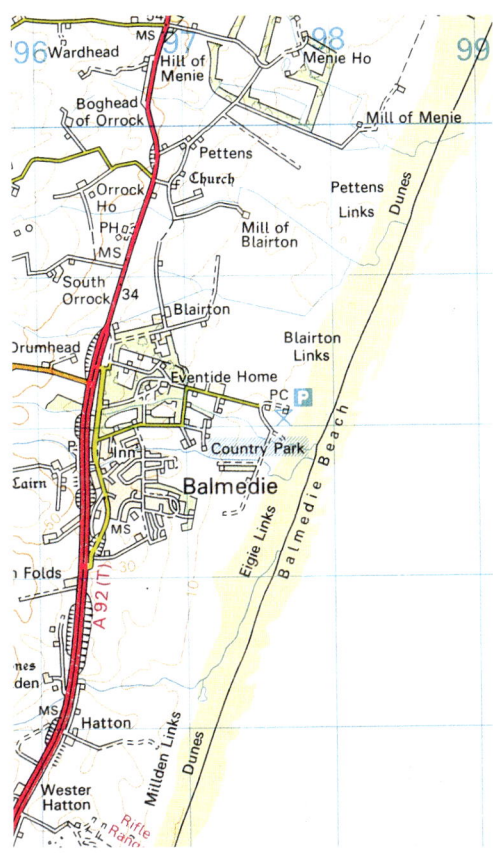

*The constantly shifting sands of Balmedie have created Sahara-like landscapes where majestic dunes sometimes reach 20 metres in height.*

Scotland

# ST COMBS
## Grampian
*OS Ref:NK056632*

Extensive sandy beaches are a feature of this part of the coast, with superb dunes extending 14 kilometers south to Peterhead. North of the dune system the attractive coastal village of St Combs, which takes its name from St Colomba, is characterised by rows of 18th-century fishermen's cottages, usually set gable-end to the sea for protection.

**Water Quality**
No routine sewage discharge has been identified.

**Bathing Safety**
Safe with the usual precautions; there is no lifeguard cover.

**Litter**
The beach is cleaned regularly between April and September.

**Access**
Access is off the B9033 six kilometers south-east of Fraserburgh.

**Parking**
Parking facilities in the village.

**Food**
A number of cafés and pubs in the village sell hot and cold food and refreshments.

*Scotland*

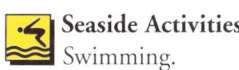

*Seemingly endless stretches of sand, backed by extensive dunes, are a feature of this stretch of the coast.*

**Seaside Activities**
Swimming.

**Wet Weather Alternatives**
Kinnaird Head lighthouse museum and leisure centre, Fraserburgh; Aden country park and heritage centre, Mintlaw.

**Wildlife and Walks**
The extensive dune system plays host to a wide range of flora and fauna. To the south, the Loch of Strathbeg is an important overwintering ground and migration staging post for wildfowl: the 2,300-acre RSPB reserve attracts over 40,000 geese, ducks and swans between September and April. Every summer many rare and unusual birds are recorded in the area. Because access to the reserve involves crossing Ministry of Defence property from the A952 road at Crimond, visitors must obtain a permit in advance from the RSPB.

**Tourist Information**
Aberdeen and Grampian Tourist Board, North Silver Street, Aberdeen, AB1 1RJ.
Tel. 01224 848848

**Track record PPPG**
Not EU designated.

*Scotland*

# ROSEHEARTY
### Grampian
*OS Ref:NJ933675*

The ancient harbour at Rosehearty is one of the longest established in Scotland, dating back to the time of the Vikings. Once a key centre for the area's fishing industry, it has now been overtaken in this role by the nearby port of Fraserburgh. Rosehearty's sandy beach nestles between the harbour to the west and rocky outcrops to the east. Visitors to the beach are greeted by picturesque views of the Moray Firth and keen walkers will be delighted to learn that the bay is linked to a network of coastal footpaths covering some 60 miles.

### Water Quality
No routine sewage discharge has been identified.

### Bathing Safety
There is no lifeguard cover.

### Litter
The beach is cleaned regularly though kelp does accumulate from time to time.

### Access
Access is via the B9031 from Fraserburgh. The beach can be reached from the road or the adjacent car park.

### Parking
There is a car park adjacent to the beach.

### Public Transport
The town is served by a regular bus service from Fraserburgh.

### Toilets
The beach has modern and well maintained toilet facilities, including those for disabled visitors.

### Food
A number of outlets sell food and refreshments in the town.

### Seaside Activities
Swimming and windsurfing.

### Wet Weather Alternatives
Kinnaird Head lighthouse museum; Fraserburgh leisure centre; Aden country park and heritage centre; Nearby Fraserburgh has a range of additional wet-weather facilities.

### Wildlife and Walks
The beach is linked to an expansive network of coastal footpaths. Many of the surrounding roads are devoid of traffic and offer superb views.

*Scotland*

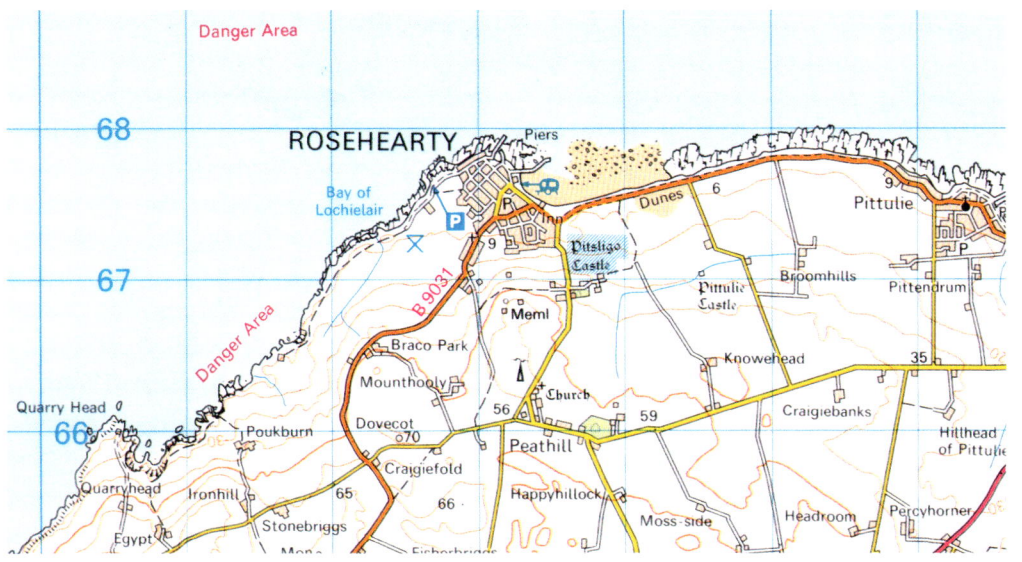

### Tourist Information
Aberdeen and Grampian Tourist Board, North Silver Street, Aberdeen, AB1 1RJ.
Tel. 01224 848848

### Track record PGGG
Not EU designated.

*Rosehearty Beach invites walkers to discover the many coastal paths in the area.*

## Scotland

| RATING | NAME | TRACK RECORD | SEWAGE OUTLET | REMARKS |
|---|---|---|---|---|
| | **BORDERS** | | | |
| P | **Eyemouth** NT945640 | PPPP | 2, both raw, both at LWM (stormwater also). | 🟨 ⛰️ Bathing can be unsafe due to currents. |
| G | **Coldingham Bay** NT918666 | PPGG | Septic tank, primary, 50, at LWM. | 🟨 Not featured due to adjacent discharge. |
| G | **Pease Sands** - EU NT794710 | PPPG | Secondary, above LWM. | FEATURED |
| | **LOTHIAN** | | | |
| G | **Thorntonloch** NT753746 | PGGG | 🟩 | FEATURED |
| ~ | **Dunglass** NT774724 | PPG~ | 🟩 | 🟨 ⛰️ |
| G | **White Sands Bay** NT710773 | PGGG | 🟩 | FEATURED |
| G | **Dunbar East** NT686786 | FPFG | Stormwater, at LWM. | 🟨 ⛰️ 🗑️ Not featured due to adjacent discharge. |
| G | **Belhaven Beach** - EU NT658786 | GGPG | Screened, 6,020, 1,200m below LWM (LSO) (stormwater also). | ⬆️ 2005 🟨 Not featured due to adjacent discharge. |
| G | **Peffersands** NT622829 | PGGG | 🟩 | 🟨 [??] Backed by dunes. |
| P | **Seacliff** NT605846 | PGGP | 🟩 | 🟨 Sheltered bay. |
| P | **Milsey Bay** - EU NT565853 | PPPP | Stormwater, at LWM. | 🟨 ⛰️ 🗑️ |
| P | **North Berwick Bay** NT553855 | FPPP | 🟩 | 🟨 ⛰️ 🗑️ |
| P | **Yellowcraig** (Broad Sands Bay) - EU NT515860 | PPPP | Raw, 900, at LWM. | ⬆️ 1997 🟨 ⛰️ 🗑️ |
| G | **Gullane** - EU NT476834 | PPGG | 🟩 | 🟨 Dunes. Not featured due to adverse reports. |
| ~ | **Gosford Sands** NT449787 | P~~~ | Raw, at LWM. | 🟨 ⛰️ |
| F | **Longniddry** NT438776 | FPFF | Primary, 4,550, at LWM (stormwater also). | ⬆️ 2005 🟨 ⛰️ 🗑️ |
| F | **Seton Sands** NT411759 | P~PF | Raw, 3,950, at LWM. | ⬆️ 1997 🟨 ⛰️ |
| ~ | **Fisherrow** NT323731 | P~~~ | Screened, stormwater only. | 🟨 🗑️ |
| P | **Portobello** NT304745 | FFPP | Stormwater at LWM beyond rocks at Joppa. | 🟨 Dangerous old wooden stands visible at low tide. |
| ~ | **Silverknowes** NT204722 | P~~~ | 🟩 | 🟨 🟫 |
| P | **Cramond** NT192771 | PPPP | 2, macerated, 3,850 (total), both below LWM. Emergency overflow above LWM. | 🟨 |
| | **FIFE** | | | |
| ~ | **Dalgety** | ~~~~ | Primary, 7,740, at LWM. | ⬆️ 🟨 ⛰️ |

🟨 Sand    🟧 Shingle    🟦 Pebbles    ⛰️ Rocks    🟫 Mud    ❓ No information supplied

# Scotland

| RATING | NAME | TRACK RECORD | SEWAGE OUTLET | REMARKS |
|---|---|---|---|---|
| P | **Aberdour**<br>Harbour<br>NT194850 | PFPP | Primary, 8,070, 1200m below LWM. | ⬆ 1998/9 🟨 ⛰ |
| P | Silversands - EU<br>NT203856 | PGGP | Primary, 750, LSO (stormwater also). | 🟨 |
| F | **Burntisland**<br>NT239858 | PFFF | Primary, 750, LSO (stormwater also). | 🟨 🗑 |
| P | **Pettycur** - EU<br>NT264862 | PPPP | 20, 20m above LWM. Septic tanks or pumped to Kinghorn. | ⬆ 🟨 🗑 |
| F | **Kinghorn**<br>NT272868 | PFPF | Primary, 1,000, LSO. | 🟨 🗑 |
| F | **Kirkcaldy Linktown**<br>NT281904 | PFFF | Stormwater. | 🟨 |
| P | **Pathhead Sands**<br>(Kirkcaldy Harbour)<br>NT292923 | PPPP | Primary, 1,960, 500m below LWM. | 🟨 Coal spoil. |
| ~ | **Leven**<br>West<br>NO386005 | F~~~ | Screened, 54,355, LSO (stormwater also). | ⬆ 1999 🟨 ⛰ |
| F | East<br>NO396014 | PFFF | Screened, 92,000, 160m below LWM. | ⬆ 1999 🟨 🗑 |
| ~ | **Lundin Links**<br>NO410022 | P~~~ | 🟩 | 🟨 To west rocks. |
| P | **Lower Largo**<br>NO417022 | PFFP | Primary, 890, 200m below LWM. | 🟨 |
| F | **Upper Largo**<br>NO427025 | PPFF | (see Lower Largo) | 🟨 |
| P | **Earlsferry** (Shell Bay)<br>NO462003 | FGGP | Primary, 1,530, 5m below LWM. | 🟨 |
| G | **Elie/Earlsferry**<br>NO485998 | PPGG | 🟩 | FEATURED |
| ~ | **Pittenweem**<br>NO550022 | P~~~ | Primary, 900 (winter), 270m below LWM. | ⛰ Fishing port. |
| P | **Anstruther**<br>NO564031 | PPPP | Stormwater, at LWM. | 🟨 ⛰ |
| G | **Roome Bay** (Crail)<br>NO618078 | PPGG | 🟩 | FEATURED |
| P | **St Andrews**<br>East<br>NO518164 | PFPP | Primary and disinfected, 200, at LWM. | ⬆ 1997 🟧 ⛰ |
| G | West Sands - EU<br>NO504175 | GGGG | Primary and disinfected, 16,000, at LWM. | ⬆ 2000/1 🟨 Not featured due to adjacent discharge. |
| P | **Tentsmuir Sands**<br>NO5024 | PPPP | Screened, stormwater. | 🟨 Dunes. |
| P | **Tayport**<br>NO463306 | FFPP | Screened, 3,430, 100m below LWM (stormwater also). | ⬆ 2000/1 🟨 |
| | **TAYSIDE** | | | |
| P | **Broughty Ferry**<br>NO469306 | PFPP | 🟩 | 🟨 |
| P | **Monifieth**<br>NO500320 | FPFP | Screened 70,000, 2km LSO. | ⬆ 2001 🟨 |

🟩 No discharge identified   ⬆ Improvements planned   ❓ Insufficient information to feature   🗑 Cleaned regularly

## Scotland

| RATING | NAME | TRACK RECORD | SEWAGE OUTLET | REMARKS |
|---|---|---|---|---|
| F | **Carnoustie** - EU NO565341 | FPPF | 🟩 | 🟨 |
| P | **Westhaven** NO574347 | PFPP | Screened, 14,000, 10m below LWM. | ⬆ 2001 🟨 |
| P | **Arbroath** - EU NO630400 | PPPP | Screened and stormwater, 25,500, 800m below LWM. | ⬆ 2001 🟨 |
| P | Victoria Park NO651410 | PPPP | Screened and stormwater, at LWM. | 🟨 |
| P | **Lunan Bay** NO6951 | PPPP | Private septic tanks. | 🟨 |
| P | **Montrose** - EU NO728579 | PGPP | 2, screened, 14,000 (total), below LWM. Others from Ferryden and Rossie Island. | 🟨 |
| | GRAMPIAN | | | |
| P | **St Cyrus** NO757648 | FPPP | Macerated, 820, at LWM. | 🟨 Saltmarsh. |
| P | **Inverberie** | ~~~P | Macerated,170m, 1,700. | 🟧 |
| G | **Stonehaven - Cowie** NO876858 | PPPG | 2, macerated/primary, 9,000 (total), LSO. | ⬆ 🟨 Not featured due to adjacent discharge. |
| P | **Stonehaven - Garron** | ~~~P | (see above) | 🟨 |
| ~ | **Muchalls** NO9092 | ~~~~ | Stormwater. | 🪨 ⛰ Bathing unsafe. |
| G | **Aberdeen** - EU NJ954072 | PPPG | (see below) | 🗑 Not featured due to adjacent discharge. |
| P | Footdee NJ958060 | PGPP | 2, raw, 8,000 (total) to harbour in tidal river. | 🟨 🗑 Located at southern point of Aberdeen beach. |
| G | **Balmedie** NJ9818 | PPGG | Secondary, LSO. | FEATURED |
| P | **Collieston** NK040285 | PPPP | Macerated, 200, at LWM. | Old fishing port. |
| P | **Cruden Bay** NK090356 | PFPP | Macerated, 2,200, at LWM. | 🟨 Dunes. |
| G | **Peterhead Lido** NK123451 | PPGG | 🟩 | FEATURED |
| G | **St Combs** NK056632 | PPPG | 🟩 | FEATURED |
| P. | **Fraserburgh** - EU NK001660 | PPPP | 12, raw, 15,690 (total). | ⬆ 2001 🟨 🗑 Dunes. |
| G | **Rosehearty** NJ933675 | PGGG | 🟩 | FEATURED |
| ~ | **Banff Bridge** | ~~~~ | 2, screened, 4,180 (total), 143m and 55m below LWM. | ⬆ 2001 🟨 |
| P | **Inverboyndie** NJ671646 | PPGP | Raw, 2,800, at LWM. | ⬆ 2001 🟨 🗑 |
| F | **Sandend Bay** NJ557662 | PPPF | Primary, 280. | 🟨 |
| P | **Cullen** - EU NJ513675 | PGGP | 2, raw, 1,500 (total), both at LWM. | ⬆ 2006 🟨 🗑 |

150  🟨 Sand   🟧 Shingle   🪨 Pebbles   ⛰ Rocks   🟫 Mud   ℹ No information supplied

# Scotland

| RATING | NAME | TRACK RECORD | SEWAGE OUTLET | REMARKS |
|---|---|---|---|---|
| ~ | **Findochty** | ~~~~ | 2, raw and macerated, 1,050 (total), at and below LWM. | ⬆ 2001 🟨 🗑 Swimmers beware rocky outcrops. |
| P | **Strathlene, Buckie** NJ448671 | P~PP | 3, all raw, 865 (total), all below LWM. | ⬆ 2001 🟨 🗑 |
| F | **Lossiemouth** East NJ240705 | PPPF | Screened, 42,700, LSO. | 🟨 🗑 Dunes. Bathing dangerous. |
| ~ | West NJ212712 | P~~~ | Stormwater. | 🟨 🗑 |
| P | Silversands | ~~GP | Preliminary treatment, 80,000, 1,200m LSO. | 🟨 |
| ~ | **Hopeman** NJ144697 | ~~~~ | Primary, 1,800, 280m below LWM. | ⬆ 2001 🟨 🗑 Dangerous rocks under the water on either side of bay. |
| ~ | **Burghead** NJ1068 | ~~~~ | Macerated, 1,440, LSO. | ⬆ 2001 🟨 🪨 🗑 |
| | HIGHLAND | | | |
| P | **Nairn** East - EU NH893574 | PPPP | Primary, 9,000, below LWM (stormwater also). | ⬆ 🟨 🗑 |
| P | Central NH883572 | PGPP | Stormwater, below LWM. | 🟨 Small area of rockpools at western end. |
| ~ | **Rosemarkie** NH7357 | P~~~ | Stormwater. | 🟨 🟧 Gravel. |
| ~ | **Cromarty** NH7867 | ~~~~ | 2, raw, 550 (total), both below LWM. | ⬆ 1998/9 🟨 |
| ~ | **Nigg Bay** | ~~~~ | 🟩 | 🟨 |
| ~ | **Portmahomack** NH915844 | ~~~~ | Secondary, 450, below LWM (stormwater also). | 🟨 Dunes. |
| P | **Dornoch** NH805890 | PGGP | 🟩 | 🟨 |
| ~ | **Sinclairs Bay,** (near Wick) ND3455 | ~~~~ | 2 raw, 3 primary, 738 (total), both near LWM. | ⬆ 1999 🟨 |
| ~ | **Duncansby Head** ND4073 | ~~~~ | 🟩 | 🟨 |
| ~ | **Dunnet Bay/Murkle Bay** ND2170 | ~~~~ | 2, 1 macerated, 1 primary, 1135 (total), at and below LWM (stormwater also). | 🟨 Dunes. |
| ~ | **Thurso** ND1168 | P~~~ | Macerated, 9,480, below LWM. | ⬆ 2005 🟨 |
| ~ | **Sandside Bay** NC9665 | ~~~~ | Macerated, 340, below LWM (stormwater also). | Dunes, rocky outcrops. |
| ~ | **Coldbackie** NC6060 | ~~~~ | 🟩 | 🟨 Dunes. |
| ~ | **Sango Bay/Balnakeil Bay** NC4068 | ~~~~ | Primary, 200, at LWM. | 🟨 Dunes. |
| ~ | **Sandwood Bay** NC2165 | ~~~~ | 🟩 | 🟨 Dunes. |

🟩 No discharge identified    ⬆ Improvements planned    ⁇ Insufficient information to feature    🗑 Cleaned regularly

## Scotland

| RATING | NAME | TRACK RECORD | SEWAGE OUTLET | REMARKS |
|---|---|---|---|---|
| ~ | Scourie NC1544 | ~~~~ | Primary, 200, at LWM. | 🟨 |
| ~ | Clashnessie Bay NC0531 | ~~~~ | 🟩 | 🟨 |
| ~ | Clachtoll NC0327 | ~~~~ | 🟩 | 🟨 |
| ~ | Achmelvich NC0524 | ~~~~ | 🟩 | 🟨 |
| ~ | Achnahaird NC0113 | ~~~~ | 🟩 | 🟨 |
| ~ | Achiltibuie NC0109 | ~~~~ | 🟩 | 🟧 |
| ~ | Gruinard Bay NG9490 | ~~~~ | 🟩 | 🟨 |
| ~ | Gairloch NG7977 | P~~~ | 4, primary, 505 (total), all below LWM (stormwater also). | ⬆ 2000 🟨 |
| ~ | Applecross NG7145 | ~~~~ | 🟩 | 🟨 |
| ~ | Coral Beaches NG2254 | ~~~~ | 🟩 | 🟨 Shells. |
| ~ | Morar NM6792 | ~~~~ | Primary, 100, below LWM. | 🟨 |
| ~ | Camusdarrach NM6691 | ~~~~ | 🟩 | 🟨 Dunes. |
| ~ | Traigh, Arisaigh NM6387 | ~~~~ | 🟩 | 🟨 Dunes. |
| ~ | Sanna Bay NM4469 | ~~~~ | 🟩 | 🟨 |
|  | STRATHCLYDE |  |  |  |
| ~ | Calgary Bay | ~~~~ | 🟩 | 🟨 Dunes. |
| ~ | Erraid | ~~~~ | 🟩 | 🟨 Dunes. |
| ~ | Kilchattan Bay | ~~~~ | Raw, below LWM. | 🟨 |
| ~ | Kames Bay | ~~~~ | 3, raw, 1100 (total), all at LWM. | 🟨 🪨 |
| ~ | Dunoon (West Bay) | ~~~~ | Raw, 1,800, below LWM. | ⬆ 2005 🟨 🪨 |
| ~ | Macrihanish | ~~~~ | Raw, 200, at LWM. | 🟨 |
| ~ | Carradale | ~~~~ | 2, raw, 500 (total), both at LWM. | |
| ~ | Helensburgh | ~~~~ | Screened and primary, 14,500, at LWM. | ⬆ 2000 🟨 🪨 |
| ~ | Portkil/Meikleross | ~~~~ | 2, primary, 168 (total), below and above LWM. | 🟨 ⛰ |
| ~ | Gourock (West Bay) | ~~~~ | 🟩 | 🟧 ⛰ |
| ~ | Lunderston Bay | ~~~~ | Private septic tanks. | 🟨 🟧 |
| ~ | Wemyss Bay | ~~~~ | 🟩 | 🟨 🟧 |

152  Sand  Shingle  Pebbles  Rocks ▪ Mud  No information supplied

*Scotland*

| RATING | NAME | TRACK RECORD | SEWAGE OUTLET | REMARKS |
|---|---|---|---|---|
| ~ | Largs | ~~~~ | Screened, 12,000, below LWM. | ⬆ 2005 🟨 ⛰ 🗑 |
| ~ | Fairlie | ~~~~ | 2, raw, 1,500 (total), at and below LWM. | ⬆ 1997 🟨 ⛰ |
| ~ | Millport | ~~~~ | 11, primary, 1,300 (total), all at LWM. | 🟨 ⛰ |
| ~ | Seamill | ~~~~ | 3, raw, 4,500 (total), all at LWM. | ⬆ 🟨 ⛰ |
| ~ | Ardrossan (Boydston) | ~~~~ | 🟩 | ⬆ 🟨 ⛰ |
| P | Saltcoats - EU NS236420 | FFPP | 🟩 | 🟨 🗑 |
| ~ | Stevenston | ~~~~ | Screened and primary, 50,000, 1,000m below LWM. | ⬆ 2000 🟨 |
| P | Irvine (Beach Park) - EU NS306377 | FFPP | 🟩 | 🟨 🗑 |
| ~ | Gailes | ~~~~ | Screened and primary, 140,000, 1,500m below LWM. | ⬆ 2000 🟨 |
| ~ | Brodick Bay | ~~~~ | 2, raw, 500 (total), above and below LWM. | 🟨 ⛰ |
| ~ | Lamlash Bay | ~~~~ | 6, raw, 1,220 (total) all at LWM. | 🟨 ⛰ |
| ~ | Whiting Bay | ~~~~ | 2, primary, 800 (total), below LWM. | 🟨 🟧 |
| ~ | Blackwaterfoot | ~~~~ | 4, primary, 200 (total). | 🟨 🟧 |
| ~ | Troon North | ~~~~ | 🟩 | 🟨 |
| P | South - EU NS321307 | PPPP | 🟩 | 🟨 |
| P | Prestwick - EU NS345262 | FFPP | 🟩 | 🟨 |
| P | Ayr - EU NS331219 | FFFP | Screened, 16,200, 140m below LWM. | ⬆ 1997 🟨 |
| ~ | Doonfoot | ~~~~ | Macerated, 13,800, 220m below LWM. | ⬆ 1997 🟨 ⛰ |
| ~ | Butlins (Heads of Ayr) | ~~~~ | Secondary, 8,500, below LWM. Private treatment plant. | 🟨 |
| ~ | Maidens | FFFF | Primary, 600, at LWM. | ⬆ 1998 🟨 ⛰ |
| F | Turnberry - EU NS200058 | ~~~~ | Primary, chlorination. | 🟨 ⛰ |
| F | Girvan - EU NX182944 | PFFF | 3, 1 screened and macerated, 2 primary, 7,000 (total), at and below LWM. | ⬆ 1997 🟨 |
| | **DUMFRIES AND GALLOWAY** | | | |
| F | Stranraer Marine Lake NX053615 | PFFF | Primary, 10,000, below LWM. | ⬆ Area affected by silt and effluent from local creamery. |

🟩 No discharge identified    ⬆ Improvements planned    ❓ Insufficient information to feature    🗑 Cleaned regularly

## Scotland

| RATING | NAME | TRACK RECORD | SEWAGE OUTLET | REMARKS |
|---|---|---|---|---|
| F | Cockle Shore NX080620 | FFFF | Creamery effluent | ☐ See above. |
| P | **Portpatrick** (Outer Harbour) NW997541 | PPFP | Primary, 600, above LWM in rocks. | ▲ |
| ~ | **Portlogan Bay** | ~~~~ | Primary, 75, 200m below HWM. | ☐ |
| P | **Drummore** NX135369 | PFFP | 2, primary, 330 (total). | ☐ |
| ~ | **Ardwell Bay** | ~~~~ | Primary, 75, 75m below HWM. | ☐ ☐ |
| P | **Sandhead** NX103501 | PPPP | Primary, 250, to river. | ☐ |
| P | **Monreith** NX359410 | PPPP | Primary, 100, above LWM to Monreithburn. | ☐ |
| P | **Mossyard** NX552518 | PPPP | ■ | ☐ |
| P | **Carrick Shore** NX575498 | PPPP | ■ | ☐ |
| P | **Brighouse Bay** NX636455 | PPPP | Secondary, 400, 30m below LWM. | ☐ |
| P | **Dhoon** NX657486 | PPPP | 2, primary, 6,200 (total). | ☐ Silt. |
| P | **Rockcliffe** NX847537 | PPPP | 2, secondary, 500 (total). | ☐ |
| P | **Sandyhills** - EU NX894553 | PFPP | Secondary, 220. | ☐ |
| P | **Southerness** | ~PPP | Primary, 3,500, 500m above LWM. | ☐ ▲ |
| F | **Powfoot** NY147654 | FFFF | Primary, 400. | Silty. |
| F | **Annan** NY198649 | FFFF | Primary, 7,900, above LWM. | ⬆ 2000 ■ Unsuitable for bathing due to deep channel and nearby sewage outfall. |

☐ Sand　▨ Shingle　▨ Pebbles　▲ Rocks　■ Mud　❓ No information supplied
■ No discharge identified　⬆ Improvements planned　⁇ Insufficient information to feature　♒ Cleaned regularly

154

# WALES

THE SAND DUNES OF ANGLESEY, THE POUNDING OF SURF ON THE LLEYN, ENDLESS KILOMETRES OF SAND, THE CLIFFS AND SECLUDED COVES OF DYFED, THE MEANDERING ESTUARIES, THE BEAUTIFUL GOWER PENINSULA – THIS IS THE COAST OF WALES. THIS CHAPTER COVERS ALL THE BATHING WATERS IN WALES IDENTIFIED UNDER THE EC BATHING WATER DIRECTIVE, AND INCLUDES MANY BEACHES NOT IDENTIFIED UNDER THE TERMS OF THE DIRECTIVE.

●

There are beaches in Wales which compare with the best anywhere in the country, and which often have the added advantage of being relatively uncrowded. Sadly, a number of individual beaches throughout the region have failed to meet even the minimum EC standard for bathing water through contamination by sewage. More positively, the problem has been recognised and is being addressed: Dwr Cymru (Welsh Water), the private company responsible for sewerage in Wales, was first among the water companies to pledge to treat all its sewage discharges to at least secondary level, with tertiary treatment where necessary. Wessex Water has now followed Dwr Cymru's lead. This work will take a long time to complete and until then there will undoubtedly be pollution blackspots around the Welsh coast; it is, however, a major step forward and the other water companies would do well to follow this lead.

Encouraging though it is, even this will not solve all Wales' pollution problems; the north and south coasts suffer from industrial effluent as well as sewage contamination. Swansea, Cardiff, Port Talbot and Newport all play a part in polluting the south coast with discharges of domestic and industrial waste. The north coast is directly affected by discharges from Merseyside and the Wirral, while Milford Haven has suffered oil pollution from the terminals and refineries that line its shore and has also had to put up with the harmful effects of antifoulants used on oil tankers. This stretch of coastline was dealt a grievous blow last year when the grounding of the tanker *Sea Empress* off Milford Haven resulted in a massive spillage of oil into the waters of Carmarthen Bay. While the worst physical signs of the pollution may have disappeared, the underlying havoc wreaked on the wildlife and habitats of this unique and important environment is still evident in 1997.

The best beaches are to be found further west, away from the centres of population – clean and unspoilt, and often wild and romantic, many are worth exploring at leisure.

# LLANDDONA - RED WHARF BAY
## Isle of Anglesey
*OS Ref: SH566812*

This is a long, wide sandy beach located at the end of Red Wharf Bay, backed by dunes and low grassland and enclosed by rounded headlands. The beach is ideal for walking, beachcombing, bathing and paddling.

### Water Quality
No routine sewage discharge has been identified.

### Bathing Safety
Bathing is generally safe, but caution should be exercised when the tide is going out. Lifebuoys are located at strategic points along the beach.

### Litter
The beach is cleaned daily during the summer by hand, and waste bins are provided. Dogs are allowed on the beach at all times.

### Access
Llanddona is signposted off the B5109 Beaumaris to Pentraeth road. The road to the beach is steep and access is via an unsurfaced bridleway. Sandy paths lead through the dunes to the beach.

### Parking
A car park has 60 parking spaces.

### Toilets
Toilet facilities are available.

### Food

A café which serves mainly light

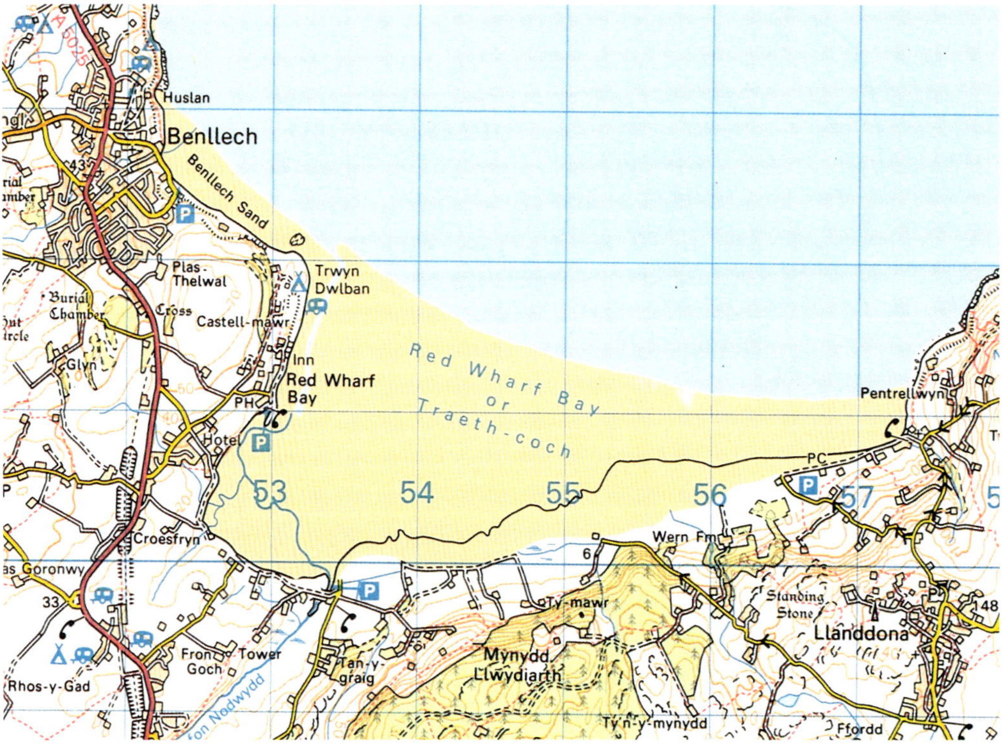

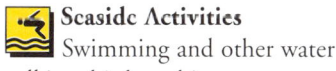

refreshments is located on the beach.

### Seaside Activities
Swimming and other water sports, walking, birdwatching.

### Wet Weather Alternatives
The fine little town of Beaumaris, with its 13th-century castle, museums and other historic buildings, is nearby. There is also a leisure centre, Anglesey Sea Zoo and a number of working farms.

### Wildlife and Walks
There are various attractive walks on offer in and around the area; for further details contact the Tourist Information Centre.

*Ten square miles of sand represent a beachcomber's delight, but walkers should beware of the incoming tide which rapidly floods the beach.*

### Tourist Information
Anglesey Tourist Information Centre, Pringles Centre, Hollyhead Road, Llanfar Pwll, Anglesey, LL61 5UJ. Tel. 01248 713173

**Track Record ~~GG**
Not EU designated.

# LLANDDWYN BEACH
## Niwbwrch, Isle of Anglesey
*OS Ref: SH403630*

With its long ribbon of sand backed by forest and five kilometres of high dunes, and its spectacular views across to the mountains of the Lleyn Peninsula dipping into the horizon, Llanddwyn is acknowledged as one of the finest beaches in Britain. A path leads through the grassy hills from the beach to the nature reserve at Llanddwyn Island.

### Water Quality
No routine sewage discharge has been identified.

### Bathing Safety
Safe, except near Abermenai Point. Lifebuoys and an emergency phone are located in the car park.

### Litter
The beach is cleaned daily during the summer. Bins are provided in the car park.

### Access
Signposted off the A4080 in Niwbwrch (Newborough); a forestry commission toll road leads to the beach.

### Parking
There is a large car park and picnic area close by the beach.

### Public Transport
The nearest railway stations are at Bodorgan and Llanfairpwll; there are buses to Niwbwrch.

### Toilets
There are toilets, including facilities for disabled visitors.

### Food
Niwbwrch has a range of stores, pubs and cafés selling hot and cold food and refreshments.

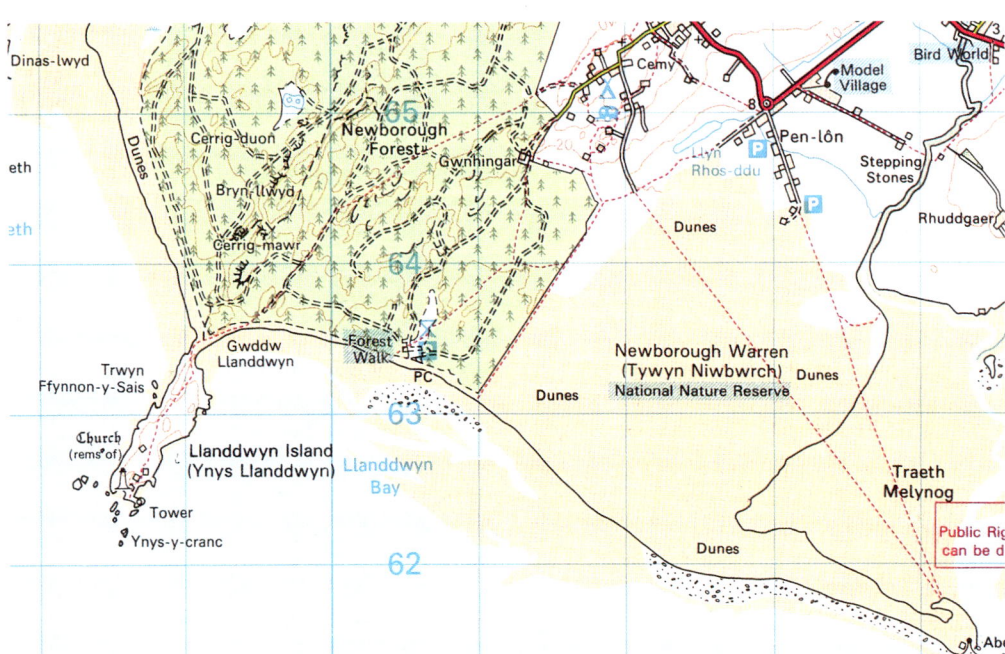

*Llanddwyn Island is easily identified by its distinctive stone lighthouse.*

### Seaside Activities
Swimming, canoeing, beachcombing and birdwatching.

### Wet Weather Alternatives
Anglesey Heritage Gallery: an introduction to the island's history, folklore and culture; Anglesey Sea Zoo; the National Trust mansion at Plas Newydd; Pili Palas (butterfly palace); Beaumaris Gaol, a model prison from Victorian times.

### Wildlife and Walks
Newborough Warren National Nature Reserve is just behind the beach and covers 1,565 acres. It can be reached via the A4080 which crosses the River Cefni at Malltraeth Pool; approach the reserve through Niwbwrch village just after the junction with the B4421.

### Tourist Information
Anglesey Tourist Information Centre, Pringles Centre, Hollyhead Road, Llanfar Pwll, Anglesey, LL61 5UJ. Tel. 01248 713173

**Track Record PGGG**
EU designated beach.

# ABERYSTWYTH NORTH and SOUTH BEACH
## Ceredigion
*OS Refs: SN583822, SN579814*

Aberystwyth is one of Wales' favourite seaside towns: it has a gently sloping sand-and-shingle beach and lies in the middle of the 70-mile-wide Cardigan Bay. The town has been a popular seaside resort since the 19th century, with its fine Edwardian promenade, medieval castle, bandstand and traditional entertainments.

### Water Quality
One outfall serving 19,132 people discharges secondary disinfected effluent.

### Bathing Safety
There is lifeguard cover between 10am and 6pm during July and August. Lifebuoys are located at regular intervals.

### Litter
The beach is cleaned daily by hand during the summer season. Dogs are banned from North Beach from May to September. Bins are provided on the beach and promenade.

### Access
Via the A487 from the north and south, and the A44 from the east. The beaches are accessible from the main road running along the front.

### Parking
There are spaces next to the beach and car parks within walking distance.

### Public Transport
The town is well served by bus and rail.

### Toilets
These include baby changing and facilities for disabled visitors.

### Food
Widely available in Aberystwyth.

### Seaside Activities
Swimming, surfing, sailing and boat trips.

### Wet Weather Alternatives
Ceredigion Museum, National Library of Wales and cliff railway.

### Wildlife and Walks
Penglais Nature Park; local walks. Contact the Tourist Information Centre for details.

### Tourist Information
Aberystwyth Tourist Information Centre, Lisburne House, Terrace Road, Aberystwyth, SY23 2AG. Tel. 01970 612125

### Track Record
**Aberystwyth - North PPPG**
**Aberystwyth - South FFPG**
Both beaches are EU designated.

# NEW QUAY, HARBOUR BEACH
## Ceredigion
*OS Ref: SN391599*

New Quay is a picturesque and internationally renowned resort village and was once a flourishing fishing port and ship building centre. Dylan Thomas lived here in the 1940s and New Quay is probably the 'cliff perched town at the far end of Wales' on which he based Llaregub in *Under Milkwood*. New Quay's Harbour Beach with its deep golden sands and secluded harbour offers both bathing and boating facilities, and is the busiest of the resort's three beaches.

### Water Quality
No routine sewage discharge has been identified.

### Bathing Safety
Bathing is generally safe. There are lifebuoys at regular intervals along the beach.

### Litter
The beach is cleaned daily by hand during the summer. Dogs are banned from the beach between May and September and litter bins are provided along its length.

### Access
The beach is located off the main road into New Quay. A path suitable for the disabled visitor leads down to the sand.

### Parking
There are car parks within walking distance of the beach.

### Toilets
These include facilities for disabled visitors.

### Food
There are food stores, pubs and cafés in New Quay.

### Seaside Activities
Swimming, sailing, mackerel fishing, boat trips.

### Wet Weather Alternatives
Contact the Tourist Information Centre for details.

### Wildlife and Walks
The Bird Hospital, south-east of the town and open to visitors on certain summer afternoons, treats oiled and injured seabirds and seals.

### Tourist Information
New Quay Tourist Information Centre, Church Road, New Quay. Tel. 01545 560865

**Track Record PPPG**
Not EU designated.

# MWNT, CARDIGAN
## Ceredigion
*OS Ref: SN193519*

The 300 metres of gently sloping sands at Mwnt Beach are fringed by folded and faulted shale and mudstone cliffs and shadowed by the imposing form of Foel-y-Mwnt, a conical hill on the headland. The tiny whitewashed Church of the Holy Cross nestles in a hollow at its foot. The only other obvious sign of man is the remnant of a lime kiln adjacent to the path down to the beach; limestone was landed in the bay and fired ready for use by the local farmers.

### Water Quality
No routine sewage discharge has been identified.

### Bathing Safety
Bathing is safe inshore, but care is required as surface currents arising from waves breaking on the headland deflect across the bay. There is an emergency phone on the cliff path.

### Litter
Dogs are banned between May and September.

### Access
Mwnt is signposted from the B4548 north of Cardigan. Lanes lead to the car park above the beach. Steps and a steep path lead down the cliff to the beach.

### Parking
There is a National Trust car park with 250 spaces.

### Toilets
These are located at the head of the steps to the beach, with facilities for the disabled visitor.

### Food
Refreshments are available from Easter to October.

### Seaside Activities
Swimming.

### Wet Weather Alternatives
Cardigan is the nearest centre.

### Wildlife and Walks
Superb National Trust cliff-top walks; a pack detailing these and other walks in the Cardigan area is available from local Tourist Information Centres. Foel-y-Mwnt hill on the headland provides good views of the bay south to Cardigan Island and the narrow rocky inlet to the north. Birdlife on the Teifi estuary and marshes (Dyfed Wildlife Trust) can be viewed from the B4546 at St Dogmaels (there is a hide at SN182458). The Welsh Wildlife Centre,

*A natural suntrap surrounded by National Trust land, this undeveloped sandy beach is easily accessible and can be very popular in summer.*

a few miles further upriver near Cilgerran, has one of Britain's largest colonies of Cetti's warblers.

### Tourist Information
Cardigan Tourist Information Centre, Theatr Mwldan, Cardigan. Tel. 01239 613230

**Track Record GGGG**
Not EU designated.

# ABERMAWR
## Pembrokeshire
*OS Ref: SM882347*

This is a remote, sandy rural beach with a pebble bank, backed by earthed cliffs. Low tide exposes tree stumps buried in the sand, the remnants of a forest drowned by a sudden flood as the ice sheet melted 8000 years ago, and perfectly preserved by the salt water.

**Water Quality**
No routine sewage discharge has been identified.

**Bathing Safety**
Bathing is generally safe, but beware of possible rip currents. There is no lifeguard cover but basic safety equipment is provided.

**Litter**
The beach is cleaned when necessary and there are daily checks during the summer.

**Access**
A road leads off the A487, the main Fishguard to St David's road, signposted to Abermawr, about 5 miles from Fishguard. There is a cliff path to the beach.

**Parking**
Parking is by the roadside.

**Toilets**
None at this beach.

**Seaside Activities**
Swimming, surfing and fishing.

**Wildlife and Walks**
Spectacular scenery from surrounding cliff walks.

**Tourist Information**
Fishguard Tourist Information Centre, 4 Hamilton Street, Fishguard, Pembrokeshire. Tel. 01348 873484

**Track Record ~~~G**
Not EU designated.

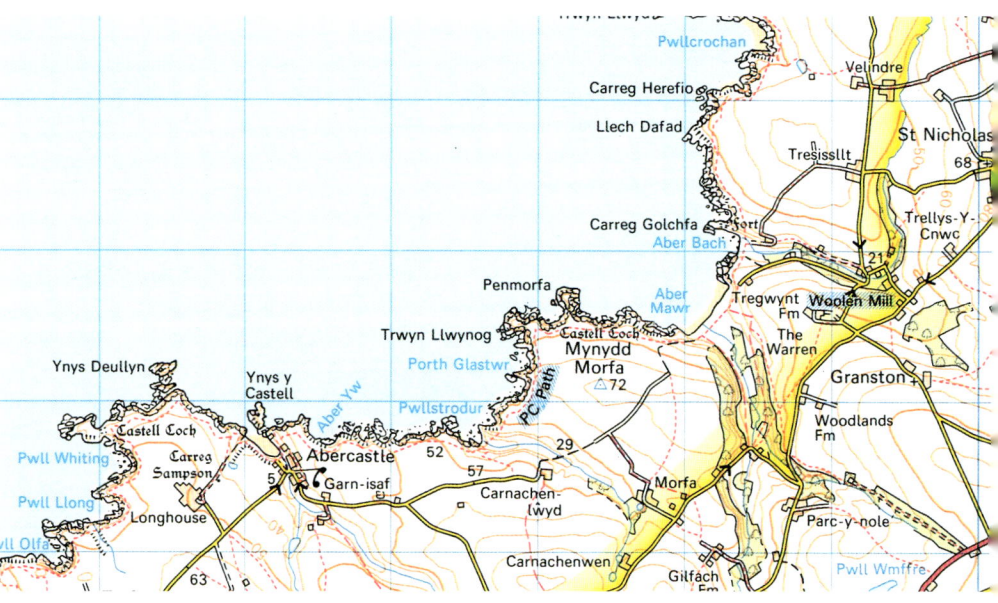

# TRAETH LLYFN
## Pembrokeshire
### OS Ref: SM802319

This is an isolated rural beach, backed by high cliffs, where even the highest of spring tides fail to completely cover the sand

### Water Quality
No routine sewage discharge has been identified.

### Bathing Safety
Beware of strong rip currents. There is no lifeguard cover, although torpedo buoys are provided.

### Litter
The beach is cleaned daily during the summer.

### Access
From the A487 Fishguard to St David's road, turn off to Portgain. A private road from Barry Island Farm ends in steep steps which lead to the beach.

### Parking
There is a private car park.

### Toilets
None at this beach.

### Seaside Activities
Swimming.

### Wet Weather Alternatives
The 19th-century harbour town of Porthgain is worth exploring.

### Wildlife and Walks
Spectacular scenery from local coastal walks.

### Tourist Information
St David's Tourist Information Centre, City Hall, St David's. Tel. 01437 720392

### Track Record ~~~G
Not EU designated.

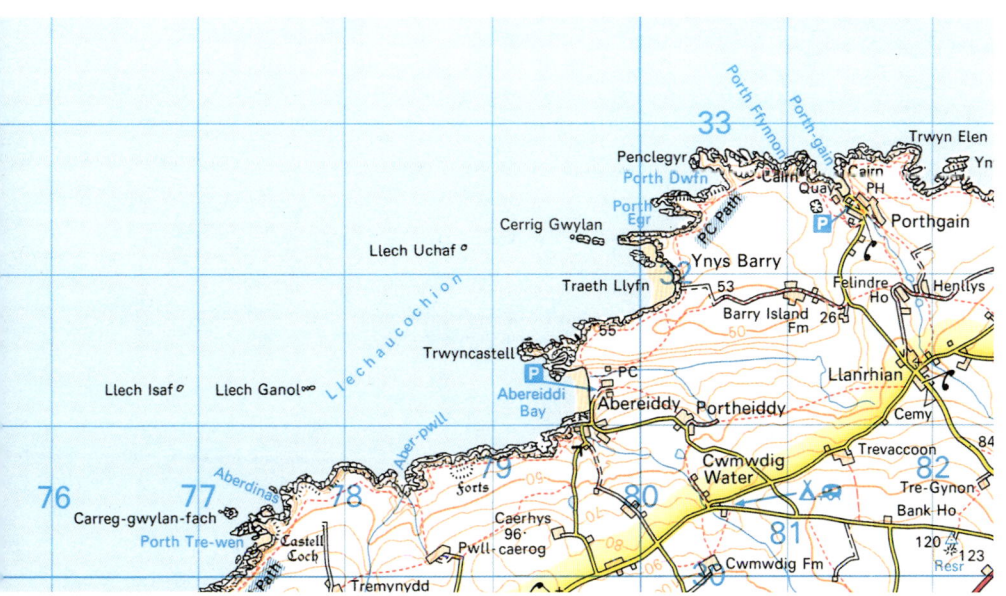

# ABEREIDDY BAY (AT SLIPWAY)
## Pembrokeshire
*OS Ref: SM795314*

Pebbles and sand made of pounded grey slate form this rural beach, and the same slate gives a brilliant deep blue colour to the water in the extraordinarily beautiful little harbour – a breached quarry – just to the north of the beach. This is a fine starting point for a coastal walk: surrounded by National Trust land there is magnificent cliff scenery at every turn and the evening sunsets are marvellously romantic.

# Wales

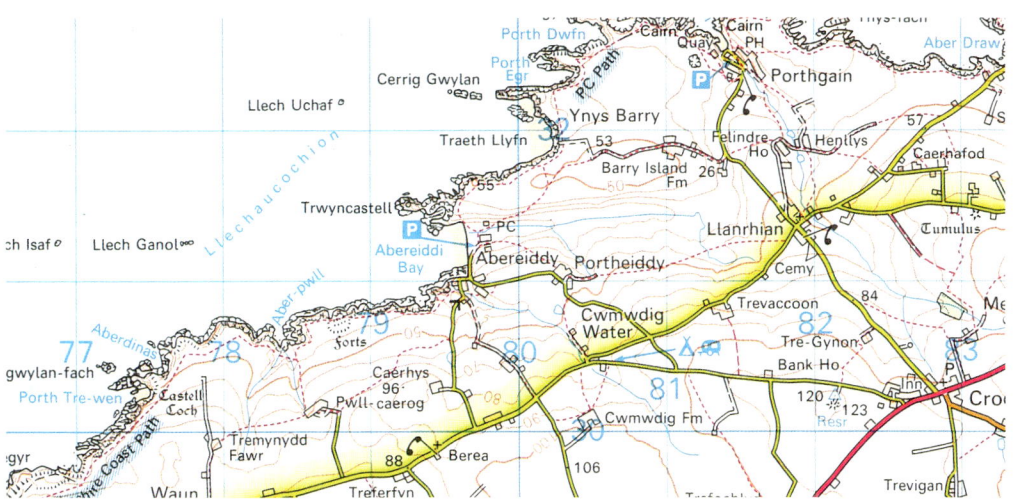

### Water Quality
No routine sewage discharge has been identified.

### Bathing Safety

This beach faces west and swells and waves can develop unexpectedly, with dangerous undercurrents and undertows. Lifesaving equipment and an emergency telephone are provided; there is no lifeguard cover.

### Litter
Cleaned daily in the summer.

### Access
From the A487 follow the lanes from Croesgoch to Abereiddy; a slipway leads down to the beach.

### Parking
Next to the beach.

### Public Transport
Buses stop at Croesgoch.

### Toilets
There are toilets behind the beach.

*Known locally as the Blue Lagoon, the harbour provides ideal anchorage for small boats.*

### Food

A mobile kiosk selling ices and teas visits the car park.

### Seaside Activities

Swimming (with caution), boating, surfing and canoeing. Care should be taken, as heavy surf can develop within 20 minutes.

### Wet Weather Alternatives

Tregwynt woollen mills with café at St Nicholas. Llangloftan cheese factory with café near Mathry. There are craft shops at Mathry and Trefin.

### Wildlife and Walks

Spectacular cliff scenery and walks from Abereiddy to Porthgain: this land is owned by the National Trust and one interesting walk explores an old industrial quarry site at Porthgain where stone was shipped to the roads in Bristol.

### Tourist Information
St David's Tourist Information Centre, City Hall, St David's. Tel. 01437 720392

**Track Record PGPG**
Not EU designated.

# WHITESANDS BAY, ST DAVID'S
## Pembrokeshire
*OS Ref: SM732271*

This wide expanse of fine white sand curving north towards the remote rocky headland of St David's is one of the best surfing beaches in the country. Open fields slope down to the shore from the imposing craggy hill Carn Llidi, which provides good walking with excellent sea views of the group of islands known as the Bishops and Clerks: the South Bishop can be identified on the far horizon by its lighthouse.

### Water Quality
No routine sewage discharge has been identified.

### Bathing Safety
Warning signs indicate where to bathe and flags indicate when it is safe: there are dangerous and unpredictable currents off parts of the beach and at some states of the tide. Lifeguards patrol the beach during the summer and first aid is available at the lifeguard centre.

### Litter
The beach is cleaned regularly; dogs are banned from May to September.

### Access
From the A487 north of St David's, signposted for Whitesands. This beach is suitable for wheelchair-users and people with mobility difficulties.

### Parking
There is a car park behind the beach with approximately 400 spaces.

### Toilets
There are toilets at the car park including facilities for the disabled.

### Food
There is a café and shop in the car park.

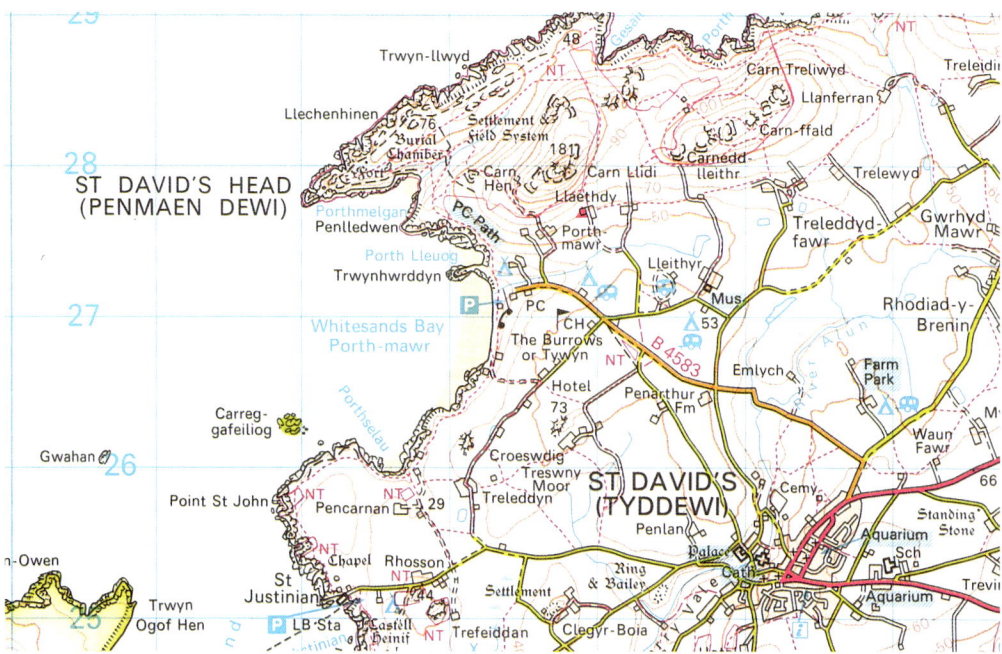

### Seaside Activities
Swimming, surfing and canoeing, with zoning of activities on busy days.

### Wet Weather Alternatives
Numerous facilities in St David's.

### Wildlife and Walks
The coast path north provides an interesting circular walk, taking in St David's Head with the remains of a fort and a burial chamber, and returning around Carn Llidi Hill. A guide describing the route is published by the Pembrokeshire Coast National Park and can be obtained at information offices locally. Ramsey Island lies just south of the bay, and boat trips from Whitesands (in

*Gorgeous sunsets framed in the wide arc of Whitesands Bay are an added attraction of this lovely beach.*

12-person inflatables between May and September) or St Justinians (daily) take you around the islands to view the seabird colonies. This area is excellent for chough, raven and peregrine.

### Tourist Information
St David's Tourist Information Centre, City Hall, St David's.
Tel. 01437 720392

**Track Record GPGG**
EU designated beach.

# NEWGALE SANDS
## Pembrokeshire
*OS Ref: SM846217*

Facing due west, Newgale Sands is a beautiful stretch of beach over 2 miles in length. The clean waters found here offer superb bathing when calm, though when storm winds blow from the Atlantic the beach is pounded by spectacular breakers. The sands are easily reached by car as, unlike a number of beaches in the area, they are served by an adjacent road. For those who enjoy walking, however, the beach is conveniently close to the Pembrokeshire Coastal Path and Coastal National Park.

### Water Quality
One outfall serving 2058 people discharges disinfected secondary treated effluent.

### Bathing Safety
Beware of large surf. Lifeguards patrol from the end of June until the first week in September.

### Litter
The beach is cleaned daily during the summer. There are no bins, so take your litter home. Dogs should not be allowed to foul the beach.

### Access
From Haverfordwest take the A487 to St David's. Newgale beach is located 7 miles along the road. Access to the beach is from the car park, over a pebble ridge which may present problems for less mobile visitors.

### Parking
Three car parks next to the beach.

### Public Transport
The nearest railway station is at Haverfordwest. A bus services runs between Haverfordwest and St David's.

### Toilets
These include facilities for the disabled visitor, and are located adjacent to the beach.

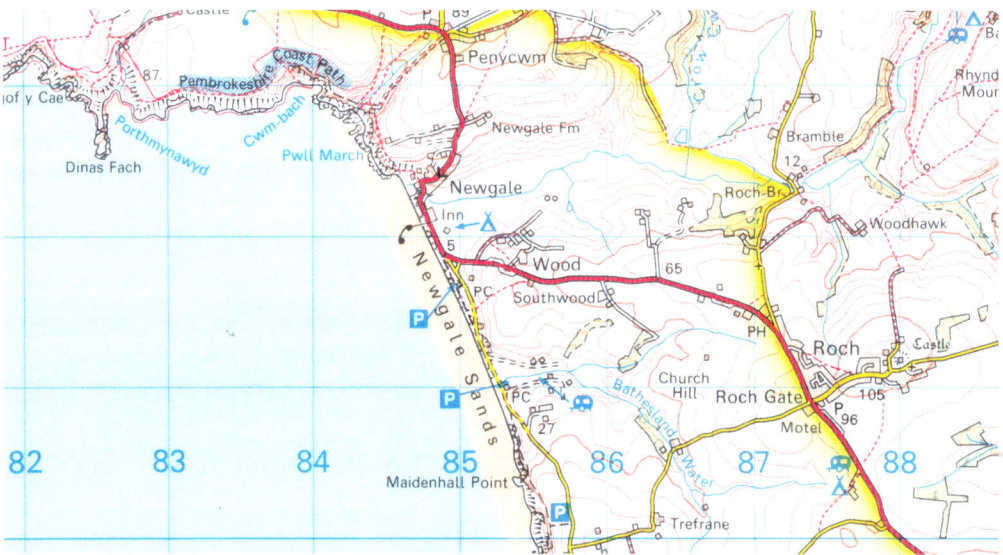

*The broad sands and clean waters of Newgale are among the most accessible in Pembrokeshire.*

### Food
Cafés at either end of the beach serve a range of hot and cold food and drink and ice-creams and teas are available from a mobile kiosk.

### Seaside Activities
This is one of the best beaches in Pembrokeshire for surfing and surf-canoeing. Equipment can be hired from Garage lifeguards on the beach.

### Wet Weather Alternatives
St David's Sea Aquarium and various other facilities in town; contact the Tourist Information Centre for details.

### Wildlife and Walks
The beach is situated within the Pembrokeshire National Park and is on the Pembrokeshire Coastal Path.

### Tourist Information
St David's Tourist Information Centre, City Hall, St David's. Tel. 01437 720392

**Track Record GPGG**
EU designated beach.

# BROAD HAVEN NORTH
## Pembrokeshire
*OS Ref: SM859138*

Flanked by the town of Broad Haven, this sandy beach faces due west, offering excellent views across St Brides Bay which looks particularly picturesque at sundown. Safe bathing means that the beach is popular with families, and the nearby town offers a number of attractions. For those with an interest in natural history, the beach also harbours a number of rock pools which are ripe for exploration and, on a larger scale, the Pembrokeshire National Park Information Centre is conveniently close at hand, from where information on guided nature walks can be obtained.

# Wales

### Water Quality
One outfall serving 1598 people discharges secondary treated effluent.

### Bathing Safety
Bathing is considered relatively safe. Large waves may accompany westerly winds. There are no lifeguards.

### Litter
The beach is cleaned daily during the summer. There is no dog ban, but owners must not let their dogs foul the beach. Recycling bins are located in the south car park.

### Access
From Haverfordwest take the B4341 signposted Broad Haven. The beach is accessed directly from the adjacent road.

### Parking
There are two car parks at either end of the beach.

### Toilets
Toilets are situated adjacent to the car parks at either end of the beach. Disabled facilities are located at the south car park.

### Food
A variety of outlets sell food within a short distance from the beach.

### Seaside Activities
Swimming, windsurfing (there is a school providing tuition), small boat sailing, surfing, canoeing.

### Wet Weather Alternatives
Sea Life Centre at St David's, Oakwood at Canaston Bridge,

*The wide sands of Broad Haven seem to run on for ever.*

Kaleidoscope at Milford Haven, ten pin bowling at Haverfordwest, karting at Wittybush Aerodrome, motor museum at Keeston.

### Wildlife and Walks
The Pembrokeshire Coast Path runs by the beach: details of organised walks can be obtained from the National Park Information Centre at Broadhaven. Natural arches and other interesting rock formations can be seen in the cliffs to the north and south. Boat trips operate from nearby Martin's Haven to Skomer Island Terrestrial and Marine Nature Reserves, which are home to a wide range of wildlife.

### Tourist Information
Haverfordwest Tourist Information Centre, 19 Old Bridge, Haverfordwest, SA61 2EZ. Tel. 01437 763110

### Track Record PFGG
EU designated beach.

# MARLOES SANDS, MARLOES
## Pembrokeshire
*OS Ref: SM780076*

Two kilometres of wide flat golden sands stretch from the imposing bulk of Gateholm Island in the north to Red Cliff and Hooper's Point in the south, backed by steep cliffs. Beds of rock laid flat on a sea bed long ago are tilted and seem to be pushing up through the sands; their jagged outlines point skyward along the length of this glorious bay. The barnacle and seaweed covered rocky outcrops testify to the fact that the whole beach disappears at high tide, and with only two access points, visitors should take care not to get cut off at the extremities of the beach by the incoming tide.

### Water Quality
No routine sewage discharge has been identified.

### Bathing Safety
Beware of currents and submerged rocks; basic lifesaving equipment is available.

### Access
From Marloes take the lane signposted Marloes Sands.

### Parking
A National Trust car park has about 50 spaces.

### Toilets
None at this beach.

### Seaside Activities
Swimming, surfing, fishing and walking.

### Wildlife and Walks
A 3 kilometre nature trail starts from the car park. To the south-west, the coast path leads to West Dale Bay and a series of secluded beaches around the Dale Peninsula that can only be reached on foot. To the north-east, the path heads to Albion Beach and Martin's Haven. Boat trips run from here to Skomer Island, a Marine Nature Reserve renowned for its seabirds, wildflowers and seal colonies.

### Tourist Information
Milford Haven Tourist Information Centre, 94 Charles Street, Milford Haven, Pembrokeshire. Tel. 01646 690866

### Track Record GGGG
Not EU designated.

*The coast path that follows the cliff top around the bay is covered in wildflowers.*

*Wales*

# MARTIN'S HAVEN
## Pembrokeshire
*OS Ref: SM761092*

Consisting of little more than a few houses and a car park, Martin's Haven is the departure point for summer trips to the islands of Skokholm and Skomer. The beach is made up of a steep shingle and pebble bank.

### Water Quality
No routine sewage discharge has been identified.

### Bathing Safety
Bathing is generally safe. There is no lifeguard cover; safety equipment and an emergency phone is available.

### Litter
The beach is cleaned daily during the summer. There are no bins, so please take your litter home.

### Access
Take the Haverfordwest to Dale road turning off for Marloes, then follow signposts for Martin's Haven. A cliff path leads down to the beach from the car park.

### Parking
Spaces for over 150 cars.

### Toilets
These include facilities for disabled visitors.

### Seaside Activities
Swimming birdwatching; boat trips to Skomer and Skokholm.

### Wildlife and Walks
An area of wilderness known as the Deer Park (NT) offers the chance to glimpse rare birds such as the chough. Skomer Island is a birdwatcher's paradise.

### Tourist Information
Milford Haven Tourist Information Centre, 94 Charles Street, Milford Haven, Pembrokeshire. Tel. 01646 690866

**Track Record ~~~G**
Not EU designated.

# SANDY HAVEN
## Pembrokeshire
*OS Ref: SM860070*

A large expanse of subtly coloured sand is revealed at this estuary beach when the tide retreats. The eastern shore of the tidal creek is occupied by a caravan and camping site.

**Water Quality**
One outfall serving 1,372 people discharges secondary treated effluent.

**Bathing Safety**
Bathing is considered generally safe, but swift currents do arise at some stages of the tide, so take note of the conditions before swimming. Torpedo buoys are located at the beach.

**Litter**
The beach is cleaned regularly by hand. There are no bins, so please take your litter home.

**Access**
The beach is accessible through the minor roads of Herbrandston and Sandy Haven. There is a slipway on to the beach, which can be used for access for the disabled visitor.

**Parking**
There is parking for over 100 cars.

**Toilets**
None at this beach.

**Seaside Activities**
Swimming and boating.

**Wildlife and Walks**
There are three circular walks around Dale and St Ishmaels (National Park); contact the Tourist Information Centre for further details.

**Tourist Information**
Milford Haven Tourist Information Centre, 94 Charles Street, Milford Haven, Pembrokeshire. Tel. 01646 690866

**Track Record ~~~G**
Not EU designated.

# WEST ANGLE BAY
## Pembrokeshire
*OS Ref: SM853033*

About a mile to the west of the village of Angle, this large beach of soft sand with gently rolling surf and rock outcrops at either end offers views of the fortifications that once protected Milford Haven.

### Water Quality
No routine sewage discharge has been identified.

### Bathing Safety
Bathing is generally safe. There is no lifeguard cover; basic safety equipment is available.

### Litter
The beach is cleaned daily by hand throughout the summer, and bins are provided. Dogs are allowed on the beach.

### Access
Take the B4320 from Pembroke to Angle; the beach is the other side of the village. There is access to the beach for the disabled visitor.

### Parking
There are spaces for over 200 cars.

### Toilets
Toilets are available.

### Seaside Activities
Swimming

### Food
There is a café adjacent to the beach.

### Wildlife and Walks
A leaflet on the local section of the Pembrokeshire Coast Path, *The Coastal Splendour Of Angle,* is available from the Tourist Information Centre.

### Tourist Information
Pembroke Tourist Information Centre, The Commons Road, Pembroke. Tel. 01646 622388

### Track Record P~PG
Not EU designated.

*Wales*

# FRESHWATER WEST
## Pembrokeshire
*OS Ref: SR882997*

Freshwater West is a long sandy beach, backed by extensive sand dunes and cliffs to the north. Seaweed used to be collected on this shore for making laver bread, a local delicacy, and a small hut once used for drying the seaweed before it was boiled still stands on the southern headland. The beach is particularly popular with surfers, but bathing is recommended for strong swimmers only, as rip currents can develop.

**Water Quality**
 No routine sewage discharge has been identified.

**Bathing Safety**
For experienced swimmers only, as strong rip currents can develop. Take note of the warning signs. There is no lifeguard cover; basic safety equipment is provided.

**Litter**
The beach is cleaned daily in the summer and once a week in the winter. Dogs are not banned, but owners are requested to prevent their animals fouling.

**Access**
Take the B4320 from Pembroke, heading for Castle Martin or Angle and follow the signposts to the beach. A board walk gives access to the beach from the south car park, and sandy paths at the north end.

**Parking**
There is parking all along the beach

**Toilets**
These include facilities for disabled visitors.

**Food**
 The occasional visit can be expected from a catering van.

**Seaside Activities**
 Swimming, surfing and fishing.

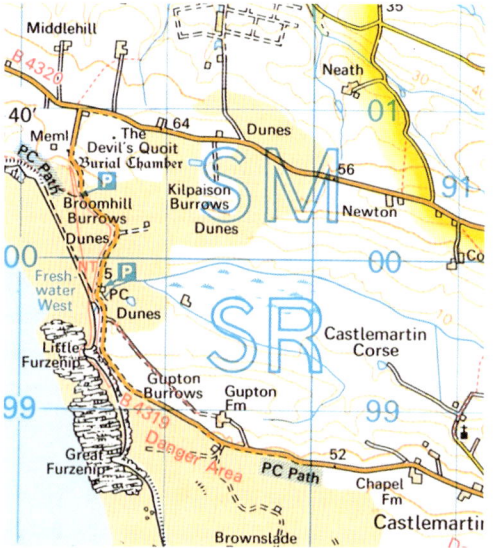

**Wet Weather Alternatives**
Pembrokeshire leisure centre, Museum of the Home and Pembroke Castle. Further details of activities can be obtained from the Tourist Information Centre.

**Wildlife and Walks**
The coast paths runs nearby; for further details contact the Tourist Information Centre.

**Tourist Information**
 Pembroke Tourist Information Centre, The Commons Road, Pembroke. Tel. 01646 622388

**Track Record ~GPG**
Not EU designated.

## BROADHAVEN – SOUTH
### Pembrokeshire
*OS Ref: SR979939*

Resting upon an impressive stretch of coast, bordered by high cliffs and facing a stack out to sea, Broadhaven South Beach rates among the most attractive rural beaches in the UK. It nestles in a deep horseshoe-shaped bay, the fine golden sands of which are bounded by rocks. As well as excellent swimming, this location also offers much for the inquisitive visitor to explore, including a series of fish ponds which lie just back from the bay and present a haven for local wildlife.

**Water Quality**
No routine sewage discharge has been identified.

**Bathing Safety**
The beach is considered safe for bathing. There is no lifeguard cover, though lifesaving equipment is on hand if needed.

**Litter**
The beach is cleaned on a regular basis.

**Access**
Signposted from the village of Bosherton, which lies just off the B4319. Access is from the car park situated above the beach, by way of a series of steep steps.

**Parking**
There is a staffed car park adjacent to the beach.

**Toilets**
Toilets are located at the head of the beach. There are no facilities for disabled

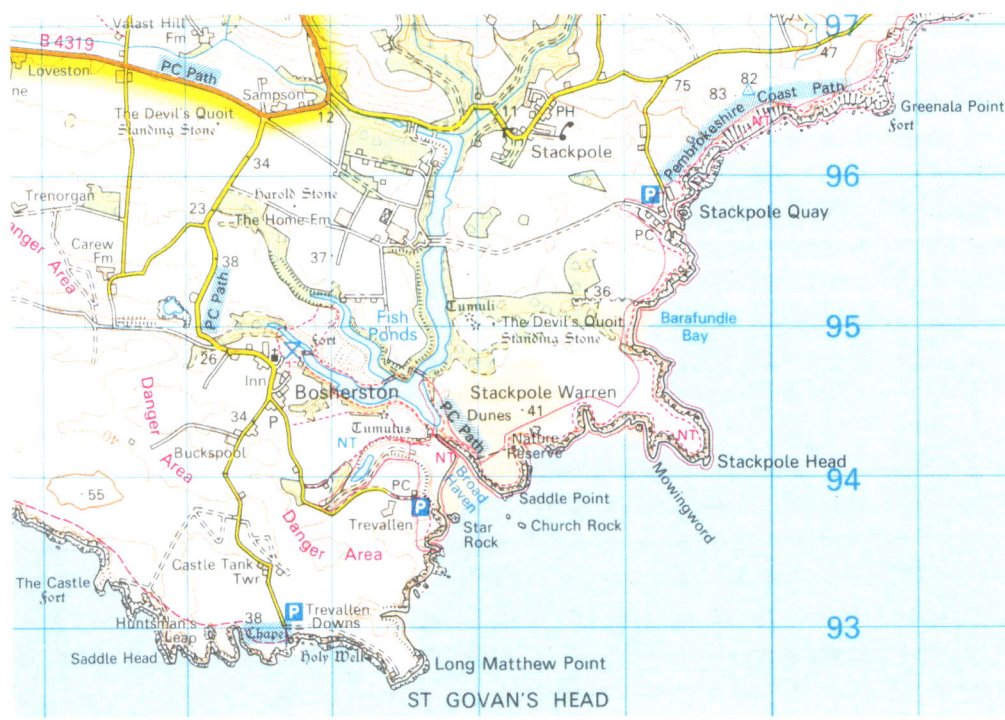

visitors at present though improvements are planned.

### Food
A stall selling snacks is located in the car park.

### Seaside Activities
Swimming and surfing.

### Wet Weather Alternatives
The beach is a short drive from Pembroke Castle. The resort town of Tenby lies ten miles to the east.

### Wildlife and Walks
The beach is owned by the National Trust and forms part of the South Pembrokeshire Heritage Coast and the Pembrokeshire Coast National Park. The beach is backed by the Pembrokeshire

*This beach figures among the very best Wales has to offer.*

Coast Path, though access through the Army's Castlemartin artillery ranges is sometimes restricted, (details from the National Park Information Centres). The Stackpole Estate (NT) lying immediately behind the bay is centred on a vast area of lily ponds created in the 18th century, and now home to a diverse range of wildlife. Tree-shaded walks enable visitors to watch the wildfowl and waders that gather there.

### Tourist Information
Pembroke Tourist Information Centre, The Commons Road, Pembroke. Tel. 01646 622388

**Track Record  P~GG**
Not EU designated.

# SKRINKLE HAVEN
## Pembrokeshire
*OS Ref: SS080973*

A steep flight of steps cut into the cliffs leads down to the sandy beach at Skrinkle Haven, easily recognised by its high rock arch which strides out into the sea.

### Water Quality
One outfall serving 2,200 people during the summer discharges disinfected secondary treated effluent below LWM at Manobier (being diverted to Tenby during 1997).

### Bathing Safety
Bathing is generally safe; torpedo buoys are provided.

### Litter
The beach is cleaned regularly by hand. There are no bins, so please take your litter home. Dogs are allowed on the beach at all times.

### Access
Take the B4586 Manobier road and follow the signs to Skrinkle. The steep stepped path down to the beach is suitable only for the fit.

### Parking
Spaces for 50 cars or more.

### Toilets
None at this beach.

### Seaside Activities
Swimming.

### Wet Weather Alternatives
Manobier Castle; a wide range of leisure activities is available in Tenby.

### Wildlife and Walks
The nearby coast path gives superb views of this dramatic coastline.

### Tourist Information
Tenby Tourist Information Centre, The Croft, Tenby.
Tel. 01834 842404

**Track Record ~~~G**
Not EU designated.

*Difficult access guarantees relative solitude at Skrinkle Haven.*

# LYDSTEP HAVEN
## Pembrokeshire
*OS Ref: SS091979*

This pretty, privately owned beach consists of sand and pebbles backed by impressive wooded cliffs at either end. At the north end of the beach there is a small sandy area exposed at low tide.

### Water Quality
No routine sewage discharge has been identified.

### Bathing Safety
Bathing is generally safe. There is no lifeguard cover, but basic safety equipment is available.

### Litter
The beach is cleaned daily by hand. Bins are provided and there is no dog ban.

### Access
Lydstep Haven is off the Tenby to Pembroke road. A toll road serving a holiday park leads to a cliff path, from where a large slipway and several sets of steps give access to the beach.

### Parking
Spaces for around 40 cars.

### Toilets
Well maintained toilet blocks include facilities for the disabled visitor.

### Food
There is a café and a fish and chip shop nearby.

### Seaside Activities
Swimming, boating, water skiing, diving and parascending.

### Wet Weather Alternatives
The 12th-century castle at Manorbier is sternly impressive, with its solid towers, gatehouse and curtain walls.

### Wildlife and Walks
See Skrinkle Haven (opposite).

### Tourist Information
Tenby Tourist Information Centre, The Croft, Tenby.
Tel. 01834 842404

**Track Record GPGG**
Not EU designated.

## PEMBREY BEACH, CEFN SIDAN
### Carmarthenshire
*OS Ref: SS400998*

Eight miles of sand edged by a belt of dunes await the visitor to this delightful beach situated within the Pembrey Country Park, which also covers the extensive grassland and forest behind the dunes. There are superb view towards the Gower Peninsula.

### Water Quality
One outfall serving 2,388 people discharges secondary treated effluent.

### Bathing Safety
Bathing is safe. Lifeguards patrol this beach near the main access point from the Spring Bank Holiday to early September.

### Litter
The beach is cleaned daily during the summer by machine. Dogs are not permitted in the central mile-long stretch of the beach between May and September. Ample litter bins are provided along the length of the beach, and a can recycler is situated at the kiosk.

### Access
The park is signposted from junction 48 of the M4, and from the A484 Carmarthen to Llanelli coast road. The park is approximately 2 miles from the A484 at Pembrey. Surfaced roads and paths lead from the main car parks. Access to the beach from the outer car parks is attained via board walks through the the dunes, not suitable for elderly or disabled visitors.

### Parking
There are several car parks behind the dunes with approximately 1,000 spaces.

### Public Transport
An hourly bus service from Llanelli and Carmarthen stops 1 mile away. Nearest rail station is Burry Port, 4 miles away, with possible taxi connection.

### Toilets
Toilets include baby changing and facilities for the disabled visitor.

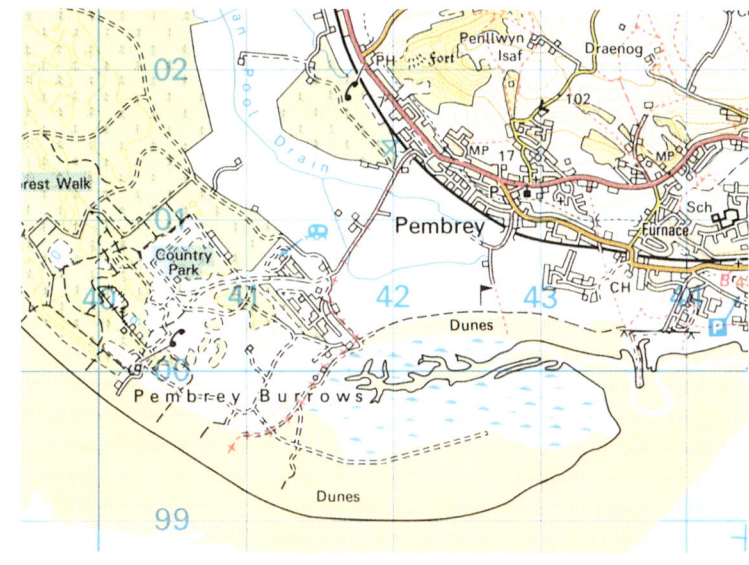

### Food
A permanent kiosk with outdoor seating serves hot and cold snacks. There is also a larger restaurant.

### Seaside Activities
Swimming, wind-surfing, land yachting, pitch and putt, railway ride, orienteering and adventure playground.

### Wet Weather Alternatives
The Wildfowl and Wetlands Trust at Penclacwydd; leisure centres and museums at Carmarthen, Llanelli and Swansea.

### Wildlife and Walks
The area is a Site of Special Scientific Interest, with 4 guided nature trails around the park (woodland, floral, beach and leisure routes). The country park also has a miniature railway, a falconry centre, a permanent orienteering course, and a two-mile off-road cycle track.

### Tourist Information
Llanelli Tourist Information Centre, Llanelli Library, Llanelli.
Tel. 01554 772020

**Track Record GPGG**
EU designated beach.

## RHOSSILI BAY, RHOSSILI
### Swansea
*OS Ref: SS414900*

A spectacular 5-kilometre sweep of golden sand fringes Rhossili Bay, from Worms Head to Burry Holms. In the north, the sands are overlooked by Rhossili Downs where grass slopes rise 200 metres above the beach. At its southern end the beach is ringed by steep cliffs which fall away northwards as down is replaced by sand dunes.

**Water Quality**
No routine sewage discharge has been identified.

**Bathing Safety**
Bathing is generally safe, but beware of rip currents during bad weather.

**Litter**
Bins are available; dogs are allowed on the beach at all times.

**Access**
From Rhossili Village a steep path and steps lead down the cliffs to the beach. A slipway makes access easier for wheelchair users and people with mobility problems.

**Parking**
There is a car park in the village.

**Public Transport**
Buses 18a and 18c from Swansea.

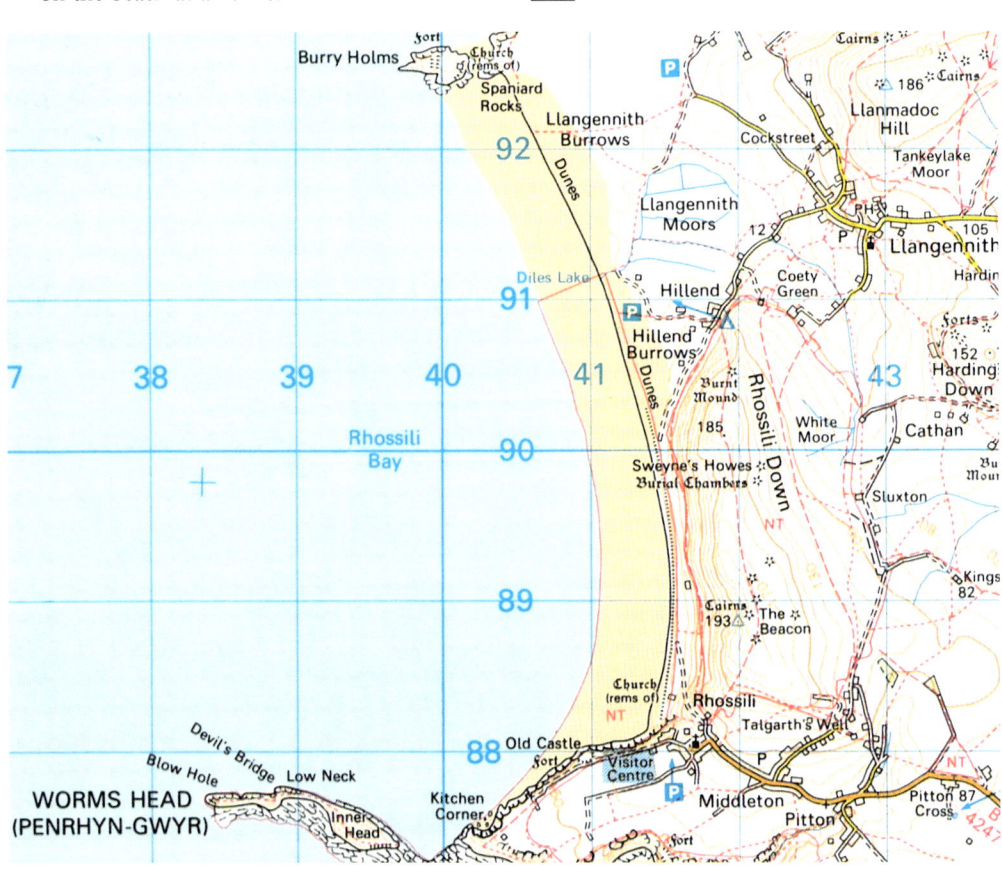

*Wales*

**Toilets**
These include facilities for disabled visitors.

**Food**
A range of refreshments is available in the village.

**Seaside Activities**
Swimming, surfing, canoeing, hang-gliding and fishing.

**Wet Weather Alternatives**
National Trust shop and information centre, Gower Heritage Centre at Parkmill.

**Wildlife and Walks**
Worms Head Island and the adjacent stretch of coast are a National Nature Reserve. The limestone cliffs are rich in flora. Nesting birds can be seen on a marked nature trail which also provides a good vantage point for sea-watching.

**Tourist Information**
Swansea Tourist Information Centre, Singleton Street, Swansea.
Tel. 01792 468321

**Track Record GGGG**
EU designated beach.

# REST & TRECCO BAYS, PORTHCAWL
## Bridgend
*OS Ref: SS800779, SS 831763*

The seaside resort town of Porthcawl once played as important a role in the industrial revolution as its near neighbours of Port Talbot and Barry. The days when coal from the valleys was exported from Porthcawl's docks are, however, long since passed and the town is now famed as a holiday centre. Together the two beaches offer all the attractions that could be expected of a traditional British holiday, along with more recently introduced pastimes, such as surfing and jet-skiing.

### Water Quality
No routine sewage discharges have been identified at either beach.

### Bathing Safety
Bathing is generally considered safe. Lifeguards patrol both beaches at weekends from May to September and daily in July and August. Flags are flown when the lifeguards are on duty.

### Litter
Both beaches are cleaned regularly. Dogs are banned from the beach between May and September, and a poop-scoop scheme is in operation.

### Access
At junction 37 of the M4, take the A4229 which leads straight to Porthcawl. At Rest Bay new access steps and a ramp have been constructed adjacent to the life guard station. For Trecco Bay, follow New Road east from the centre of the resort. The bay is reached through the caravan site, or on foot from Beach Road.

### Parking
Rest Bay car park has 500 spaces; Trecco Bay has only limited parking.

### Public Transport
The nearest rail station is at

Wales

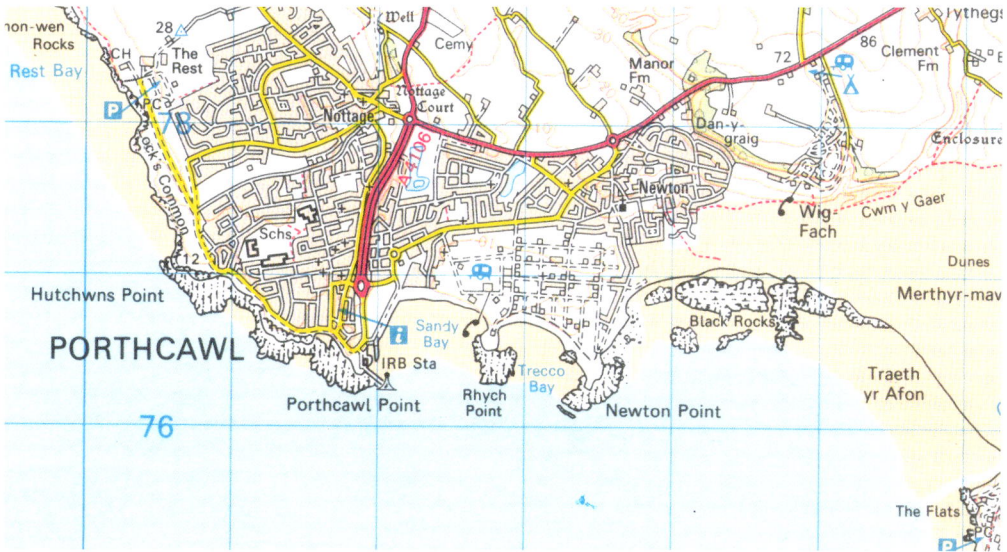

Bridgend. A regular bus service operates from Bridgend to Porthcawl.

### Toilets
Toilet facilities are next to the beaches.

### Food
There are a number of cafés, pubs and take-aways within a short distance of both beaches.

### Seaside Activities
Swimming, surfing, surf-skiing, canoeing, fishing, windsurfing and children's amusements.

### Wet Weather Alternatives
Porthcawl funfair, Grand Pavilion, swimming pool and various facilities based at the Trecco Bay Caravan Park, including ten-pin bowling, swimming pool, gym, sauna and amusements.

### Wildlife and Walks
The bays are on the Glamorgan Heritage Coast. Kenfig National Nature Reserve lies 3 miles to the west of the town.

### Tourist Information
Porthcawl Tourist Information Centre, John Street, Porthcawl. Tel. 01656 772211 or 01656 786639

**Track Record**
**Rest Bay PPPG**
**Trecco Bay PPGG**
Both beaches are EU designated.

*Clean seas, attractive sands and all the fun of a seaside resort can be found at Porthcawl.*

# Wales

| RATING | NAME | TRACK RECORD | SEWAGE OUTLET | REMARKS |
|---|---|---|---|---|
| | **FLINTSHIRE** | | | |
| P | **Point of Ayr Lighthouse** SJ121859 | ~PPP | 🟩 | 🟨 |
| | **DENBIGHSHIRE** | | | |
| P | **Prestatyn Central** – EU SJ054839 | PFPP | Screened, macerated, 33,066, 1440m below HWST. | ⬆ 1999 🟨 |
| P | **Gronant** SJ095848 | ~~~P | (see Prestatyn) | 🟨 |
| P | **Ffrith** SJ046836 | ~~PP | (see Prestatyn) | 🟨 |
| P | **Rhyl** – EU SJ002826 | FPFP | Secondary, 9,866, from Rhuddlan. | 🟨 Bathing unsafe at river mouth. |
| G | **Kinmel Bay (Sandy Cove)** – EU SH978866 | PPPG | Primary, 70,258, 4060m below HWST. | ⬆ 1998 🟨 🟧 Dunes. Bathing unsafe at river mouth. Not featured due to adjacent discharge. |
| P | **Abergele** Towyn SH965796 | ~~PP | (see Kinmel Bay – Sandy Cove) | 🟨 🟧 |
| P | Pensarn SH944790 | ~PPP | (see Kinmel Bay – Sandy Cove) | 🟨 At low tide. |
| P | **Llanddulas** SH907788 | ~PPP | Screened, macerated, 2,927, 220m below HWST, Diversion to Kimnel Bay during 1997. | ⬆ 1999 🟨 🟧 |
| P | **Colwyn Bay** – EU SH858791 | FPPP | (see Colwyn Bay – opp. Rhos Abbey Hotel) | 🟨 |
| P | Marine Road SH852793 | ~~~P | (see Colwyn Bay – opp. Rhos Abbey Hotel) | 🟨 🟧 |
| P | End of Cayley Prom SH849796 | FPPP | (see Colwyn Bay – opp. Rhos Abbey Hotel) | 🟨 |
| P | Opposite Rhos Abbey Hotel SH845804 | PPPP | Raw, 30,691, 600m below HWST. | ⬆ 1999 🟨 |
| | **CONWY** | | | |
| F | **Penrhyn Bay** SH831816 | FFFF | Screened, macerated 4,164, 210m below HWST. | ⬆ 1999 🟨 🟧 |
| P | **Llandudno** North Shore – EU SH786823 | FPGP | 🟩 | 🟨 🟧 🗑 |
| P | West Shore – EU SH765816 | FFPP | Screened, macerated, 33,507, LSO. | ⬆ 1999 🟨 🟧 🗑 |
| F | **Deganwy** (North) SH775794 | FFPF | ❓ | ⬆ 1999 🟧 |
| P | **Conwy Morfa** SH757788 | ~~~P | (see Llandudno – West Shore) | 🟨 |
| P | **Penmaenmawr** Conwy Bay – EU SH717768 | PPFP | 7,747 below HWST, nr. Dwygyfylchi. | ⬆ 1999 🟨 🟧 |
| P | **Llanfairfechan** SH680758 | FPPP | Secondary, 4,359. | ⬆ 2001 🟨 Dangerous tidal currents. |

🟨 Sand  🟧 Shingle  Pebbles  Rocks  Mud  ❓ No information supplied

*Wales*

| RATING | NAME | TRACK RECORD | SEWAGE OUTLET | REMARKS |
|---|---|---|---|---|
| | **ISLE OF ANGLESEY** | | | |
| P | **Beaumaris** SH607759 | ~~~P | Primary, 4,100, 308m below HWST. | 🟨 🟧 Bathing dangerous. |
| G | **Llanddona** – Red Wharf Bay SH566812 | ~~GG | 🟩 | FEATURED |
| P | **Benllech** – EU SH526825 | PPPP | 6,784. | ⬆ 1997 🟨 |
| ~ | **Craig Dwlan (Benllech)** SH531821 | ~~~~ | (see Benllech) | 🟨 |
| P | **Moelfre (Treath Lligwy)** SH497872 | ~~~P | 🟩 | 🟧 |
| P | **Porth Eileen** SH476929 | ~~~P | 3,001, Llaneilian 65m below HWST, preliminary treatment only. | 🟨 |
| F | **Amlwch (Bull Bay)** SH427945 | ~~~F | Raw, 3,847, 43m below HWST. | ⬆ 2005 🟨 |
| P | **Cemaes Bay** SH373936 | ~~~P | Raw, 2,126, 45m below HWST. | ⬆ 1999 🟨 |
| P | **Church Bay** SH299894 | ~~~P | 🟩 | 🟨 🟧 |
| P | **Porth Tywyn Mawr (Sandy Bay)** SH286851 | ~~~P | 🟩 | 🟨 |
| ~ | **Newry Beach** – Holyhead SH242833 | ~~~~ | 5, 3 raw, 9,768 (raw), 36 – 200m below HWST. | ⬆ Docks area. |
| P | **Trearddur Bay** – EU SH255789 | PPPP | Screened and disinfected, 10,120, 125m below HWST. | ⬆ 2005 🟨 ⛰ 🗑 |
| P | **Porth Dafarch** SH233799 | ~~~P | 🟩 | |
| P | **Borth Wen** – Rhoscolyn SH273750 | ~~~P | 🟩 | 🟨 |
| P | **Silver Bay** – Rhoscolyn SH292751 | ~~~P | 🟩 | |
| P | **Porth Nobla** SH328712 | ~~~P | 🟩 | |
| G | **Rhosneiger** (Traeth Crigyll) – EU SH323722 | GGGG | Raw, 4,381, 580m below HWST | ⬆ 1999 🟨 Backed by dunes, not featured due to adjacent discharge. |
| ~ | **Traeth Llydan** (Broad Beach) SH323721 | ~~~~ | (see Traeth Crigyll) | Dunes. |
| P | **Aberffraw Bay** SH352676 | ~~~P | Screened, macerated, 1,372, 91m below HWST. | ⬆ 1998 🟨 Extensive dune system. |
| G | **Llanddwyn Beach, Niwbwrch** SH403630 – EU | PGGG | 🟩 | FEATURED |
| ~ | **St George's Pier**, Menai Bridge | ~~~~ | 4 raw, 1 screened and macerated, 1,197 (raw), 230. | ⬆ 1997 |

🟩 No discharge identified   ⬆ Improvements planned   ⁇ Insufficient information to feature   🗑 Cleaned regularly

# Wales

| RATING | NAME | TRACK RECORD | SEWAGE OUTLET | REMARKS |
|---|---|---|---|---|
| | **GWYNEDD** | | | |
| F | **Menai Straits**<br>Porth Dinorwic Sailing Club<br>SH521673 | ~PPF | 3, raw, 630 (total), at or below HWST. | ⬆ 1997 Strong currents in Menai Straits make bathing unsafe. |
| P | Plas Menai<br>SH502661 | ~PPP | 🟩 | |
| G | **Dinas Dinlle** – EU<br>SH434566 | GGGG | 🟩 | ❓ 🟨 🩶 🗑 |
| ~ | **Pontllyfni**<br>SH430528 | ~~~~ | Secondary, 941. | 🟨 |
| ~ | **Trefor**<br>SH378474 | ~~~~ | Raw, 564, 100m below HWST. | ⬆ 1999 🟨 🟧 |
| P | **Porth Nefyn**<br>SH301408 | ~PPP | Macerated, 2,877, 109m below HWST. | ⬆ 1999 🟨 |
| ~ | **Morfa Nefyn**<br>SH282409 | ~~~~ | 🟩 | 🟨 ⛰ |
| ~ | **Porth Dinllaen** | P~~~ | 🟩 | 🟨 ⛰ |
| ~ | **Rhos-y-Llan**<br>SH230376 | ~~~~ | Primary, 164. | 🟨 |
| ~ | **Traeth Penllech** | ~~~~ | 🟩 | 🟨 |
| ~ | **Porth Colman**<br>SH201344 | ~~~~ | 🟩 | Rockpools. |
| ~ | **Porth Iago**<br>SH167316 | ~~~~ | 🟩 | 🟨 |
| ~ | **Porthor** (Oer)<br>SH166300 | ~~~~ | Primary, from public toilets. | 🟨 ⛰ |
| P | **Aberdaron Beach**<br>SH173262 | ~PPP | 1,654, tertiary treatment. | 🟨 |
| G | **Porth Neigwl Beach**<br>SH280263 | ~GGG | Primary, 77. | 🟨 Edged by steep cliffs, not featured due to adjacent discharge. |
| P | **Abersoch** – EU<br>SH316277 | PPGP | Secondary, 19,023, feasibility study in progress for further improvements. | ⬆ 🟨 |
| P | **Llanbedrog**<br>SH333314 | ~FPP | Macerated, 860, 365m below HWST, sewage flow to be diverted to Pwlheli. | ⬆ 🟨 |
| G | **Pwllheli** – EU<br>SH371340 | PGPG | Macerated, 6,973, 91m below HWST. | ⬆ 1997 🟨 Fast currents, not featured due to adjacent discharge. |
| P | **Morfa Aberech**<br>SH404359 | ~GGP | 🟩 | 🟨 Backed by dunes. |
| ~ | **Afon Wen**<br>SH443368 | ~~~~ | Macerated, 589, 317m below HWST, tidal tank. | ⬆ 2005 🟨 🟧 |
| P | **Criccieth Beach**<br>SH498376 | ~~~P | (see Criccieth Beach East) | 🩶 |
| P | **Criccieth Beach** (East) – EU<br>SH503380 | PFPP | Secondary, disinfection, 3,556. | 🟨 🟧 |

192 | 🟨 Sand | 🟧 Shingle | 🩶 Pebbles | ⛰ Rocks | 🟫 Mud | ❓ No information supplied

# Wales

| RATING | NAME | TRACK RECORD | SEWAGE OUTLET | REMARKS |
|---|---|---|---|---|
| G | **Black Rock Sands** (Morfa Bychan) – EU SH542359 | PFPG | (see Morfa Bychan) | 🟨 Not featured due to adjacent discharge. |
| ~ | **Morfa Bychan** | ~~~~ | 2, macerated, primary, 9,015 (total) 2300m below HWST. | ⬆ 1998 🟨 🗑 Do not bathe at south-east end. |
| F | **Carreg Wen** SH560370 | ~~~F | (see Morfa Bychan) | |
| F | **Harlech** – EU SH567314 | PPFF | Primary, 3,716, feasibility study in progress for further improvements. | ⬆ 🟨 |
| P | **Llandanwg** – EU SH566281 | PPPP | Primary, 3,004, feasibility study in progress for further improvements. | ⬆ 🟨 ⛰ Bathing unsafe at low tide. |
| G | **Tal-y-Bont** – EU SH577211 | P~GG | 🟩 | ⁇ 🟨 Dunes. |
| ~ | **Llanaber** (Dyffryn) | ~~~~ | 9,491, tertiary treatment. | 🟨 |
| G | **Barmouth** – EU SH608159 | PGGG | Screened, 3,582, LSO, 1,480m below HWST. | 🟨 Not featured due to adjacent discharge. |
| G | **Fairbourne** – EU SH609130 | PGPG | 1,076, 350m below HWST. Preliminary treatment only. | ⬆ 2004 🟨 🟧 Not featured due to adjacent discharge. |
| ~ | **Llwyngwril** SH588101 | ~~~~ | Raw, 871, 118m below HWST, feasibility study in progress for further improvements. | ⬆ 2005 🟨 🟧 |
| G | **Tywyn** – EU SH576003 | PPPG | Secondary, disinfection, 18,000. | 🟨 ⁇ |
| F | **Aberdyfi** – EU SN607958 | FPFF | Sewage flow transferred to Tywyn in 1996. | 🟨 |
| | **CEREDIGION** | | | |
| P | **Ynys Las** North SN603928 | ~GPP | Discharge to River Dyfi. | 🟨 Dunes 🗑 |
| P | East Tywyni, Ynys Las Estuary SN612943 | ~PFP | 🟩 | 🟨 |
| P | **Borth** – EU SN606901 | PGFP | 🟩 | 🟨 |
| P | **Clarach Bay** North of River SN585840 | FFPP | Secondary, 39, discharges into Nant Clarach from caravan sites. | 🟨 🟧 |
| ~ | South of River SN586837 | FFP~ | (see Clarach Bay - North of River) | 🟨 🟧 🗑 |
| G | **Aberystwyth** North – EU SN583822 | PPPG | (see Aberystwyth – South) | FEATURED |
| ~ | Harbour SN580810 | FFF~ | (see Aberystwyth - South) | |
| G | South – EU SN579814 | FFPG | Secondary, disinfected, 19,132. | FEATURED |

🟩 No discharge identified  ⬆ Improvements planned  ⁇ Insufficient information to feature  🗑 Cleaned regularly

## Wales

| RATING | NAME | TRACK RECORD | SEWAGE OUTLET | REMARKS |
|---|---|---|---|---|
| G | Tanybwlch Beach SN579806 | PFGG | 🟩 | 🟨 Not featured due to unsuitable bathing. |
| G | Morfa Bychan (Slipway) SN565774 | PPPG | Primary, discharges from caravan site. | ⛰ Not featured due to adjacent discharge. |
| P | Llanrhystud (South) SN523691 | PGGP | Secondary, 1,182. | 🟨 🟨 Safe bathing. |
| F | Llansantffraid SN508671 | PPPF | Primary, 1,085. | 🟨 🟨 |
| F | Llanon – (Slipway) SN506667 | PPPF | (see Llansantffraid) | 🟨 |
| P | Aberarth SN476638 | PPGP | Primary, 743. | 🟨 |
|  | **Aberaeron** |  |  |  |
| F | North of Groynes SN462635 | ~~~F | (see Aberaeron – Fourth Groyne) | 🟨 |
| P | South Beach SN453628 | PFGP | (see Aberaeron - Fourth Groyne) | 🟨 🟦 |
| F | North Beach SN455635 | ~~~F | (see Aberaeron - Fourth Groyne) | 🟨 |
| P | Fouth Groyne - North Harbour SN455634 | PFPP | Raw, 3,414, 350m below HWST. | ⬆ 1997 🟨 🟨 |
| F | Little Quay (Cei Bach) SN410598 | PFFF | 🟩 | 🟨 |
| P | Gilfach yr Halen SN435613 | ~~~P | 🟩 |  |
| P | Llanina SN403598 | ~~~P | Primary, 6,856, feasibility study in progress for further improvements. | ⬆ |
|  | **New Quay** |  |  |  |
| P | Treath y Dolau SN390602 | PPPP | 🟩 | 🟨 |
| G | Harbour SN391599 | PPPG | 🟩 | FEATURED |
| G | Traeth Gwyn – EU SN398597 | PFPG | Primary, 6,856, LSO off Llanina Point, feasibility study in progress for further improvements. | ⬆ 🟨 🗑 Not featured due to adjacent discharge. |
| P | Cwmtydu SN355576 | PPPP | 🟩 | 🟨 🟨 Toilet and parking facilities. |
| P | Cil Borth SN311544 | ~~~P | (see Llangranog) |  |
| P | Llangranog – EU SN310543 | PFPP | Secondary, 269, below LWM. | ⬆ 1997 🟨 🟨 🗑 |
| P | Penbryn SN291524 | P~PP | 🟩 | 🟨 🗑 |
| P | Tresaith – EU SN278517 | PPPP | Macerated, 221, 365m below HWST, diversion to Aberporth by summer 1997. | ⬆ 1997 🟨 🗑 |
|  | **Aberporth** |  |  |  |
| P | Traeth-y-Dyffryn SN259517 | PFPP | (see Aberporth – Slip) | 🟨 |

🟨 Sand　🟧 Shingle　🟦 Pebbles　⛰ Rocks　🟫 Mud　❓ No information supplied

*Wales*

| RATING | NAME | TRACK RECORD | SEWAGE OUTLET | REMARKS |
|---|---|---|---|---|
| P | Slip – EU SN258517 | PFGP | Primary, disinfected, 2,709, membrane filtration. | 🟨 Rockpools at low tide. |
| G | **Mwnt, Cardigan** SN193519 | GGGG | 🟩 | FEATURED |
| P | **Patch** SN162485 | ~~~P | 🟩 | |
| F | **Gwbert-on-Sea** at Craig y Gwert SN159503 | ~FFF | Secondary, discharges from caravan park and private properties. | 🟧 |
| | PEMBROKESHIRE | | | |
| F | **St Dogmaels Slipway** SN164469 | ~~~F | 🟩 | |
| | **Poppit Sands** | | | |
| P | West – EU SN152489 | PPGP | (see Poppit Sands – East) | 🟨 Dunes 🗑 Bathing dangerous in river. |
| P | East SN156492 | PPGP | Secondary, 7,618, feasibility study in progress for further improvements. | 🟨 Bathing safe only where the lifeguard indicates. |
| | **Newport Sands** | | | |
| P | North – EU SN053407 | FFPP | Macerated, 2,397, 500m below HWST. | ⬆ 1998 🟨 |
| P | South SN052405 | PGGP | (see Newport Sands – North) | 🟨 Dunes, bathing safe in centre of beach. |
| F | **Newport Car Park Slip** SN015397 | ~~~F | (see Newport Sands – North) | |
| P | **Cwm yr Eglws** SN015401 | ~~~P | 🟩 | |
| F | **Pwll Gwaelod** SN003399 | PFPF | Secondary, 882. | 🟨 |
| P | **Goodwick Harbour** – South SM948381 | PPPP | (see Goodwick Beach) | 🟨 |
| F | **Goodwick Beach** SM949379 | PFPF | 3, 2 raw, 4,200 (raw), 3,328, 60m and 470m below HWST. | ⬆ 1997 🟨 Ferry terminal. |
| G | **Abermawr** SM882317 | ~~~G | 🟩 | FEATURED |
| F | **Abercastle** | ~~~F | 🟩 | |
| G | **Traeth Llyfn** SM802319 | ~~~G | 🟩 | FEATURED |
| G | **Abereiddy Bay** (At Slipway) SM795314 | PGPG | 🟩 | FEATURED |
| G | **Whitesands Bay** – St David's – EU SM732271 | GPGG | 🟩 | FEATURED |
| P | **Caerfai Bay** SM760243 | GPGP | 🟩 | 🟨 🗑 Small rugged cove. |
| P | **Porthglais** SM742238 | ~~~P | 🟩 | |
| G | **Newgale Sands** – EU SM846217 | GPGG | Secondary, disinfection, 2,058. | FEATURED |

🟩 No discharge identified  ⬆ Improvements planned  [??] Insufficient information to feature  🗑 Cleaned regularly

## Wales

| RATING | NAME | TRACK RECORD | SEWAGE OUTLET | REMARKS |
|---|---|---|---|---|
| G | **Broad Haven** – North – EU SM861138 | PFGG | Secondary, 1,598, feasibilty study in progress for further improvements. | FEATURED |
| P | **Nolton Haven** SM857184 | ~~~P | | |
| P | **Little Haven** SM856129 | ~~PP | | Small cove with slipway. |
| P | **St Brides Haven** SM801109 | ~~PP | | |
| G | **Marloes Sands** – Marloes SM780076 | GGGG | | FEATURED |
| ~ | **Musselwick Sands** | ~~~~ | | Cliff-backed cove – possible to get cut off by the rising tide. |
| P | **Dale** SM813058 | PGPP | Macerated, 362. | 1998 |
| G | **Sandy Haven** SM860070 | ~~~G | Secondary, 1,372. | FEATURED |
| P | **Milford Beach** SM888055 | ~~PP | Secondary, 13,720, 545m below HWST, nr Hakin Point. | Near refinery town. |
| ~ | **Neyland Slip** SM967047 | ~~~~ | Primary, 3,697, feasibility study in progress for further improvements. | |
| G | **West Angle Bay** SM853033 | P~PG | | FEATURED |
| G | **Martin's Haven** SM761092 | ~~~G | | FEATURED |
| G | **Freshwater West** SR882997 | ~GPG | | FEATURED |
| G | **Broadhaven** (South) SR979939 | P~GG | | FEATURED |
| P | **Barafundle Bay** SR992950 | GPGP | | Backed by trees and dunes. |
| P | **Freshwater East** SS018976 | PGGP | Screened, 600, off Trewent Point, feasibility study in progress for further improvements. | |
| P | **Manorbier** SS058974 | GPGP | Secondary, disinfected, 2,200 (summer), below LWM, diverted to Tenby during 1997. | |
| G | **Skrinkle** SS080973 | ~~~G | (see Manorbier) | FEATURED |
| G | **Lydstep Haven** SS091979 | GPGG | | FEATURED |
| P | **Tenby** South – EU SS132998 | PPPP | Screened, macerated, 15,819, 2,800m below HWST. | 1997 Backed by vegetated sand dunes. |
| G | North – EU SN134008 | PPPG | (see Tenby South) | Not featured due to adjacent discharge. |

Sand   Shingle   Pebbles   Rocks   Mud   No information supplied

# Wales

| RATING | NAME | TRACK RECORD | SEWAGE OUTLET | REMARKS |
|---|---|---|---|---|
| P | **Saundersfoot** Beach – EU SN141047 | PFPP | 🟩 Diverted to Tenby. | 🟨 🟫 🗑 |
| P | **Coppet Hall** SN143053 | ~~~P | 🟩 | 🟨 |
| P | **Wiseman's Bridge** SN144060 | ~~~P | 🟩 | 🟨 |
| P | **Amroth – EU** SN167068 | PPGP | Secondary, 447, to stream, feasibility study in progress for further improvements. | 🟨 ⬜ 🗑 |
| | CARMARTHENSHIRE | | | |
| P | **Pendine Sands – EU** SN238074 | GPPP | Secondary, 6,685. | 🟨 Dunes. |
| F | **Llanstephan & Twyi Estuary** SN355107 | ~~~F | 2, primary, secondary 21,438 (total). | 🟨 |
| F | **Ferryside** SN363104 | ~~~F | Secondary, 902. | 🟨 |
| F | **St Ishmael** (Kidwelly) SN365075 | ~~~F | Secondary, 3,764. | |
| G | **Pembrey Beach** (Cefn Sidan) – EU SS400998 | GPGG | Secondary, 2,388. | FEATURED |
| F | **Burry Port Beach East** SN446002 | PFFF | Secondary, 5,877. | ⬆ 1997 🟫 Industrial, bathing unsafe. |
| F | **Llanelli Beach** (Fourth Groyne) SS496995 | FFFF | Secondary, 28,813. | ⬆ 1997 🟨 |
| | SWANSEA | | | |
| P | **Broughton Bay** SS419930 | PFPP | 🟩 | 🟨 Dunes, bathing unsafe at LWM and ebbing tide due to currents. |
| G | **Rhossili Bay, Rhossili** – EU SS414900 | GGGG | 🟩 | FEATURED |
| ~ | **Fall Bay** SS419871 | ~~~~ | Primary 328. | ⬆ 1997 🟨 |
| ~ | **Mewslade** | ~~~~ | (see Fall Bay) | 🟨 |
| G | **Port Eynon** – EU SS472848 | GGGG | Primary, 596, feasibility study in progress for further improvements. | 🟨 Backed by dunes, not featured due to adjacent discharge. |
| G | **Oxwich Bay** – EU SS507862 | GGGG | Secondary, 288. | ⬆ 1998 🟨 Backed by dunes, not featured due to adverse reports. |
| P | **Three Cliffs Bay** SS535876 | P~~P | (see Oxwich Bay) | 🟨 |
| P | **Southgate** (Pwlldu Bay) SS576870 | ~~~P | Secondary, 2,097. | ⬆ 1997 ⛰ |
| F | **Brandy Cove** | ~~~F | Secondary, 3,707. | ⬆ 1997 🟨 ⬜ ⛰ |
| P | **Caswell Bay** – EU SS591874 | PPPP | (see Mumbles Head | 🟨 🗑 |

🟩 No discharge identified   ⬆ Improvements planned   ⁇ Insufficient information to feature   🗑 Cleaned regularly

# Wales

| RATING | NAME | TRACK RECORD | SEWAGE OUTLET | REMARKS |
|---|---|---|---|---|
| P | **Langland Bay**<br>West<br>SS606871 | FPPP | (see Mumbles Head) | Sand |
| P | **Limeslade Bay** – EU<br>SS625870 | PFPP | (see Mumbles Head) | Some sand. |
| P | **Bracelet Bay** – EU<br>SS630871 | PPPP | (see Mumbles Head) | Rockpools. |
| | **Swansea Bay** | | | |
| F | The Mumbles – EU<br>SS644921 | FPFF | (see Mumbles Head) | Sand |
| F | County Hall<br>SS653921 | ~~~F | (see Mumbles Head) | |
| F | Opposite Black Pill Rock<br>SS632898 | FFPF | (see Mumbles Head) | Sand |
| F | At Slip | ~~~F | (see Mumbles Head) | |
| F | Sketty Lane<br>SS639914 | ~~~F | (see Mumbles Head) | |
| P | Knab Rock<br>SS625878 | PPPP | (see Mumbles Head) | Sand |
| P | Mumbles Head<br>SS632874 | FPFP | Screened, 149,419 731m below HWST. | Sand |
| | **NEATH & PORT TALBOT** | | | |
| | **Jersey Marine,** | | | |
| P | West<br>SS704925 | ~~~P | Raw, 1,200, diversion to Swansea Bay during 1997. | |
| P | Central<br>SS704925 | P~PP | (see Jersey Marine – West) | |
| P | East<br>SS709916 | ~~~P | (see Jersey Marine – West) | |
| ~ | **Baglan** (Neath) | ~~~~ | Screened, 130,693, 2,343m below HWST. | ⬆ 2000 Sand |
| ~ | **Afan** (Port Talbot) | ~~~~ | Screened, 45,623, 2,946m below HWST. | ⬆ 2000 Sand |
| P | **Aberafan** – EU<br>SS739896 | PPPP | 11, raw, 23,747 (total). | ⬆ 2000 |
| P | East<br>SS744889 | PPPP | (see Afan) | |
| ~ | Margam Sands<br>(opposite steel works) | ~~~~ | (see Afan) | Sand |
| | **BRIDGEND** | | | |
| | **Porthcawl** | | | |
| G | Rest Bay – EU<br>SS800779 | PPPG | | FEATURED |
| P | Sandy Bay – EU<br>SS824765 | PPGP | | Sand Rocks |
| G | Trecco Bay – EU<br>SS831763 | PPGG | | FEATURED |
| P | **Newton Bay** (Newton Point)<br>SS838766 | ~FPP | | Sand Rocks Bathing prohibited at Newton Point. |

Sand | Shingle | Pebbles | Rocks | Mud | No information supplied

*Wales*

| RATING | NAME | TRACK RECORD | SEWAGE OUTLET | REMARKS |
|---|---|---|---|---|
| | VALE OF GLAMORGAN | | | |
| ~ | **Ogmore-by-Sea** SS858752 | ~~~~ | Secondary, 157,517, to river, sewage pumped to Penybont and treated. | 🟨 ⛰️ Bathing prohibited near estuary. |
| P | **Southerndown** (Dunraven Bay) – EU SS884729 | PPPP | 🟩 | 🟨 ⛰️ Bathing prohibited off headland. |
| ~ | **Nash Point** SS905698 | ~~~~ | 🟩 | ⛰️ Bathing unsafe. |
| ~ | **Tresilian Bay** | ~~~~ | (see Llantwit Major) | ⛰️ Bathing unsafe. |
| P | **Llantwit Major Beach** SS955673 | PPFP | Macerated, 9,016, 150m below HWST. | ⬆️ 1998 🟨 ♻️ |
| P | **Limpert Bay** ST019662 | ~PPP | 9,881, 20m below HWST, preliminary treatment only. | 🟨 🟫 ⛰️ |
| P | **Fontygary Bay** (Rhoose) ST052657 | PPPP | (see Cold Knap Beach) | ⛰️ Difficult current. |
| P | **Barry**     Watch House Bay ST106662 | PFPP | (see Cold Knap Beach) | 🟫 |
| ~ |     Little Island Bay | P~~~ | (see Cold Knap Beach) | 🟫 |
| P | **Cold Knap Beach** – EU ST096664 | PPPP | 1 raw, 3 screened, 46,838, 192 – 2,013m below HWST, preliminary treatment only. | ⬆️ 2002 🟨 🗑️ |
| P | **Whitmore Bay** – Central – EU ST114662 | PFPP | (see Cold Knap Beach) | 🟨 🟫 🟤 |
| P | **St Mary's Well Bay** ST178673 | PFPP | (see Cold Knap Beach) | 🟫 ⛰️ |
| F | **Jacksons Bay** – EU ST122665 | PFPF | (see Cold Knap Beach) | 🟨 Unstable rocks at the back of the beach. |
| F | **Bendricks Beach** ST136670 | ~~~F | (see Cold Knap Beach) | |
| F | **Penarth** ST189708 | FFFF | 4, all raw, 22,874 (total), 25 – 213m below HWST. | ⬆️ 2002 🟫 ⛰️ Bathing dangerous. |

🟩 No discharge identified   ⬆️ Improvements planned    Insufficient information to feature   🗑️ Cleaned regularly

*The Giant's Causeway, near Portrush in County Antrim, draws thousands of visitors every year.*

# Northern Ireland

THE NORTHERN IRISH COASTLINE REMAINS A WELL-KEPT SECRET TO MANY OF THOSE OUTSIDE THE PROVINCE: IT IS BEAUTIFUL AND VARIED BEYOND IMAGINATION, RANGING FROM CLIFFS POUNDED BY THE ATLANTIC ON THE NORTH COAST OF ANTRIM TO THE SOFT, WOODED SHORES OF CARLINGFORD LOUGH IN THE SOUTH.

●

Its most famous feature, the Giant's Causeway, is one of the wonders of the natural world, consisting of huge polygonal basalt columns that disappear into the waves like a stairway to the depths. Legend has it that the Causeway was built by the giant Finn MacCool to speed his path to Scotland, and, sure enough, a similar rock formation juts out of the sea off the Scottish island of Staffa. Its actual origins are a little more mundane – the result of the slow cooling of volcanic outpourings of basalt to form the regular crystalline structure we see today. Since 1987, the Giant's Causeway has been listed as a World Heritage Site.

The whole coast has a rich and varied geology, with a succession of bays and rugged headlands. The high plateau of Antrim is broken by steep glens and coastal cliffs with spectacular views to Scotland and the Isles, while further south, the Irish Sea bites deep into the gentle hills of County Down. The beaches, too, span the range of variation, from the long sandy strands of County Londonderry with their popular holiday resorts, to the nine glens of Antrim, each with a little beach nestling at its mouth, providing a quieter alternative.

Most of the Province's sea loughs are rich in wildlife and scenery. Strangford Lough is particularly beautiful, with hundreds of kilometres of coastline and over 120 islands. In early summer seals come to Strangford to give birth to their young and a wide variety of birdlife is found there all year round. Both the uniqueness and diversity of the Lough's wildlife have resulted in it recently being designated a Statutory Marine Nature Reserve. Visitors to Northern Ireland will find their own reward in the dramatic seascapes and lovely beaches, still largely undiscovered, and some with excellent water quality.

# MAGILLIGAN STRAND, BENONE
## Co. Londonderry
*OS Ref:C723362*

This seven-mile stretch of firm flat golden sand backed by dunes was Northern Ireland's first European Blue Flag and 'Premier' Seaside Award Beach. At the eastern end cliffs rise to 700ft, themselves shadowed by a heather clad plateau and the Binevenagh mountains.

### Water Quality
No routine sewage discharge has been identified.

### Bathing Safety
Bathing is safe where indicated; lifeguards are on duty in July and August, between 10am and 6pm.

### Litter
Litter bins are provided. The beach is cleaned mechanically at least twice daily from Easter to the end of September. Dogs are banned in designated areas between May and September. Dog hygiene bins are located on the beach and access routes. A poop-scoop scheme operates throughout the year.

### Access
Benone Beach is situated 12 miles north of Limavady on the A2 seacoast road. There is an access road direct to the beach off the A2. Vehicle access to the beach is via a concrete ramp and pedestrian walkways lead from the Benone Tourist Complex and car park.

### Parking
There is free parking on the beach; a car park nearby has space for 24 vehicles, 4 of which are reserved for disabled drivers.

### Public Transport
Ulsterbus runs a daily service throughout the year and special services in the high season.

### Toilets
These include facilities for disabled visitors.

### Food
A coffee shop opens at the Benone Tourist Complex between June and September.

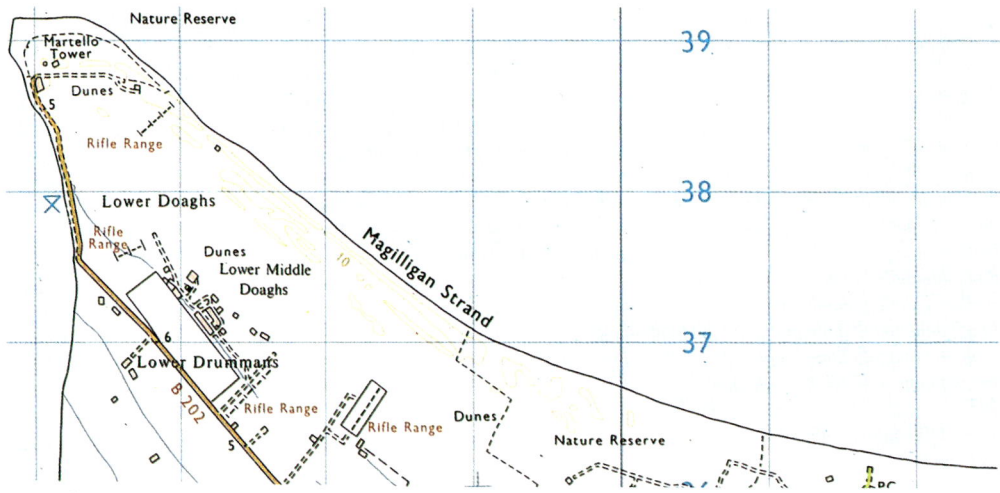

*Northern Ireland*

**Seaside Activities**
Swimming, surfing, fishing, canoeing; beach club.

**Wet Weather Alternatives**
Roe Valley Recreation Centre, Limavady and Benone Tourist Complex.

**Wildlife and Walks**
Visitors will enjoy a rich diversity of plant and birdlife at Benone: a series of organised walks take place during the summer to explore the flora and fauna in Benone and Umbra Nature Reserves. There

*Should you prefer to get away from the sandcastles and games, miles of sand offer solitude with only the sea and the sky for company.*

is excellent hillwalking further inland.

**Tourist Information**
Benone Tourist Complex, Benone, Limavady, Co. Londonderry
Tel. 015047 50555

**Track Record PGPG**
EU designated beach.

# TYRELLA BEACH, CLOUGH
## Co. Down
*OS Ref:J476357*

This enclosed beach/dune complex in Dundrum Bay comprises four kilometres of wide sandy beach backed by 25 hectares of mature dune conservation area. The clean, shallow water and safe bathing make it a very popular beach with families, especially on sunny Sundays in the holiday season; at other times it can be blissfully quiet. There are six golf courses within a 20-kilometre radius of the beach, including the excellent Royal County Down.

### Water Quality
No routine sewage discharge has been identified.

### Bathing Safety
Bathing is considered safe at all times. A warden service operates during the summer season, and at weekends from 9am to 5pm between Easter and September.

### Litter
In the summer season the beach is cleaned daily by hand and twice weekly by machine; in winter it is cleaned as required. Some marine debris is washed up on shore. Dogs are banned from the car-free zone during the main summer season. Litter bins are provided.

### Access
The beach is signposted off the A2 Killough-Clough road.

### Parking
Cars may park to the right of the access road; to the left is a car-free zone.

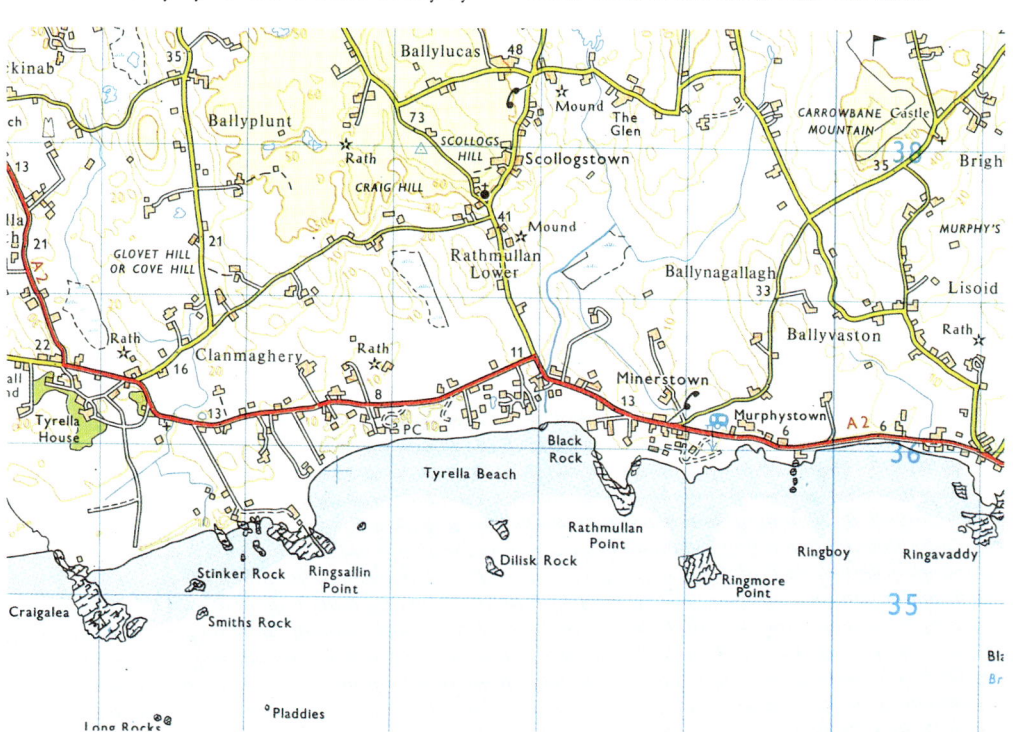

*Northern Ireland*

### Toilets
A modern, well maintained block includes baby changing and facilities for the disabled visitor

### Food
Food vans visit the beach during the summer. There is also a beach shop near the entrance point.

### Seaside Activities
Swimming, organised activities for children, beach games, guided walks and nature walks.

### Wet Weather Alternatives
Down County Museum, Delamont country park, Seaforde butterfly house, Castleward House and Gardens (National Trust). There is an information post next to the amenity block.

*A tractor follows its twice weekly track pulling a mechanical litter-picker the length of the beach.*

### Wildlife and Walks
The dunes at Tyrella are designated as a conservation area and are easily accessible via boardwalk paths. Close by is Delamont Country Park, on the shores of Strangford Lough, one of the UK's three statutory Marine Nature Reserves. The Ulster Way goes past the beach before threading its way through the beautiful Mourne Mountains to the south.

### Tourist Information
Downpatrick Tourist Information Centre, 74 Market Street, Downpatrick.   Tel. 01396 612233

**Track Record GGGG**
EU designated beach.

# NEWCASTLE
## Co. Down
### OS Ref:J384318

Lying upon the shores of Dundrum Bay, Newcastle is far the best known coastal resort town on the South Down coast. The beach itself is formed by a large expanse of sand which gives way to isolated outcrops of stone towards its southern reaches. It is backed by a promenade, providing easy access to a range of facilities typical of such a holiday centre. Newcastle also makes an ideal base for exploring the unspoilt countryside of southern County Down, including the Mountains of Mourne and Strangford Lough.

### Water Quality
One sewage outfall serving 20,000 people discharges secondary treated effluent.

### Bathing Safety
Bathing is considered safe. There is no lifeguard cover, but an R.N.L.I. station is nearby.

### Litter
The beach is cleaned daily during the summer and litter bins are provided. Dog owners are advised that a poop-scoop scheme is in operation.

### Access
The main road serving Newcastle from Belfast is the A24. The beach is easily reached from the adjacent promenade and harbour area.

### Parking
A number of car parks along the length of the beach provide ample parking. Cars are not allowed on the beach itself.

### Public Transport
A bus service operates from Belfast via Downpatrick.

### Toilets
There are several toilets situated in the car parks and along the promenade, in easy reach of the beach. A number of these include facilities for the disabled visitor.

### Food
There are various hotels, restaurants and cafés situated along the main road running parallel to the beach.

*Northern Ireland*

*Newcastle's sandy beach is set against the majestic Mourne Mountains.*

**Seaside Activities**
Swimming, windsurfing and the Tropicana Leisure Complex.

**Wet Weather Alternatives**
There are various attractions in the Newcastle area. Full details are available from tourist information offices.

**Wildlife and Walks**
Only a short drive away is the Murlough Nature Reserve, a dune system owned by the National Trust and teaming with interesting wildlife and vegetation. Newcastle also provides an excellent access point for gentle hikes or more serious hill-walking in the Mournes. Strangford Lough lies just to the north-east of town.

**Tourist Information**
Newcastle Tourist Information Centre, Central Promenade, Newcastle, Co. Down
Tel. 013967 22222

**Track Record PPGG**
EU designated beach.

# NICHOLSON'S STRAND, KILKEEL
## Co. Down
*OS Ref:J282109*

An Area of Outstanding Natural Beauty adjoining a Site of Special Scientific Interest, Nicholson's Strand stretches along the eastern shores of the entrance to Carlingford Lough. The south-facing sand and shingle beach is backed by dunes and has the Mourne Mountains as a backdrop. There are good views across the Lough to Ballagan Point and away down the coast beyond Dundalk Bay.

**Water Quality**
No routine sewage discharge has been identified.

**Bathing Safety**
Generally safe.

**Access**
A road leads to the car park a short walk from the beach.

**Parking**
There is a car park near the beach with 150 spaces.

**Toilets**
There are clearly marked public conveniences.

**Food**
A hotel, cafés and shops provide a range of refreshments.

**Seaside Activities**
Swimming, windsurfing, diving, water-skiing and fishing.

**Wet Weather Alternatives**
Annalong Cornmill and Marine Park.

**Wildlife and Walks**
The Mourne Mountains provide some of the best walking in Britain. The Silent Valley Reserve – a 14,000 million-litre reservoir set among the peaks with fine parkland on the approaches to the dam – is just north of Kilkeel. Slieve Donard, at

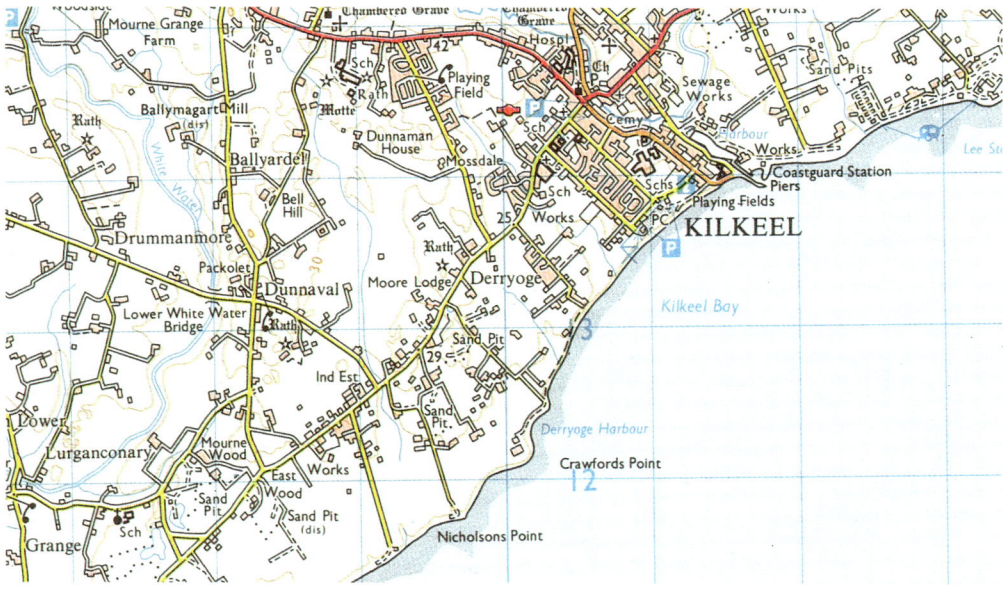

*Northern Ireland*

850m Northern Ireland's highest peak, rises just to the north-east.

**Tourist Information**
Kilkeel Tourist Information Centre, 6 Newcastle Street, Kilkeel, BT34 4AF. Tel. 016937 62525

*The south-facing sand and shingle beach at Nicholson's Strand attracts visitors throughout the season.*

**Track Record GGGG**
EU designated beach.

## Northern Ireland

| RATING | NAME | TRACK RECORD | SEWAGE OUTLET | REMARKS |
|---|---|---|---|---|
| | CO. LONDONDERRY | | | |
| G | **Benone** Magilligan Strand - EU C723362 | PGPG | 🟩 | FEATURED |
| P | **Castlerock** - EU C777364 | PPFP | | ⬆ 2000 |
| P | **Portstewart** (The Strand) - EU C808367 | PGGP | 🟩 | Dunes |
| | CO. ANTRIM | | | |
| G | **Portrush** Mill Strand - EU C856406 | GGGG | Macerated, 17,700 (sum), 5,600 (win), at LWM. | ⬆ Not featured due to adjacent discharge. |
| P | Curran Strand - EU C863406 | PGGP | 🟩 | |
| G | **Ballycastle** - EU D123412 | PGPG | Macerated, 7,500, (sum), 4,050 (win), 30m below LWM. | ⬆ Not featured due to adjacent discharge. |
| P | **Cushendun** D250327 | ~~~P | Raw, 700, below LWM, also stormwater. | ⬆ 2002 |
| G | **Carnlough** D286174 | ~~~G | Primary, 2,000, below LWM, also stormwater. | ⬆ 2002 Not featured due to adjacent discharge. |
| G | **Glenarm** D309156 | ~~~G | Primary, 840, below LWM, also stormwater. | Not featured due to adjacent discharge. |
| ~ | **White Park Bay** | ~~~~ | 🟩 | |
| P | **Browns Bay** - EU D436028 | PPGP | 🟩 | |
| | CO. DOWN | | | |
| G | **Helen's Bay** - EU J460829 | PPPG | 1,600, at LWM, tidal tank. | ⬆ Not featured due to adjacent discharge. |
| G | **Crawfordsburn** (Bangor) - EU J467826 | FGGG | Secondary, 1,200, to stream. | ⬆ ⁇ Not featured |
| G | **Ballyholme** - EU J518824 | PPPG | Stormwater, to stream. | ⬆ Not featured due to adjacent discharge. |
| G | **Groomsport** - EU J455836 | PGGG | Screened, 40,000, at LWM. | ⬆ Not featured due to adjacent discharge. |
| G | **Millisle** - EU J601755 | PPGG | Primary, 1,000, at LWM. | ⬆ Not featured due to adjacent discharge. |
| P | **Ballywater** J630682 | ~~~P | Primary, 2,000, below LWM, also stormwater. | |
| P | **Ballyherbert** J659632 | ~~~P | Primary, 1,200, below LWM, also stormwater. | |
| ~ | **Murlough** | ~~~~ | Secondary, 1,000, above LWM. | ⬆ |
| G | **Tyrella Beach** (Clough) - EU J476357 | GGGG | 🟩 | FEATURED |
| G | **Newcastle** - EU J384318 | PPGG | Secondary, 20,000, below LWM. | FEATURED |

Sand   Shingle   Pebbles   Rocks   Mud   ❓ No information supplied

*Northern Ireland*

| RATING | NAME | TRACK RECORD | SEWAGE OUTLET | REMARKS |
|---|---|---|---|---|
| G | **Nicholson's Strand** (Kilkeel) - EU J282109 | GGGG | ■ | FEATURED |
| G | **Cranfield Bay** - EU J268105 | PFGG | Screened, 2,200 (sum), 200 (win), 410m below LWM. | ☐ ⌑ Not featured due to adjacent discharge. |

■ No discharge identified  ⬆ Improvements planned  ⁇ Insufficient information to feature  ⌑ Cleaned regularly

*The magnificent 13th-century Mont Orgueil Castle overlooks the harbour at Grouville (p236).*

# The Channel Islands

The most southerly land of the British Isles, lying just 23 kilometres off the coast of France, the Channel Islands have had continuing political links with Britain since the time of William of Normandy, who ruled over them before he conquered England. Of the five main islands, only Jersey and Guernsey are covered in the guide. Alderney, Sark and Herm are fabulous places to visit, however, each with its own distinct way of life, and caring little about the outside world.

●

Sark is only 5 kilometres long and has been a feudal state governed by a Seigneur since 1565. Travel on the island is by bicycle, tractor or horse-drawn carriage – cars are not allowed. Herm is even smaller than Sark, measuring only 2.5 kilometres long by 1 kilometre wide. Shell Beach at the north of Herm consists of millions of tiny shells, some from as far away as the Gulf of Mexico. There are, not surprisingly, no cars or roads on the island, but Herm is home to around 100 of the famous Guernsey cows.

Guernsey itself is a picturesque island with spectacular cliffs and sandy beaches. Its capital, St Peter Port, is known as one of the finest harbour towns in Europe. The island has its own government, issues its own coins and stamps, but is subject to the Crown, and although the islanders are English-speaking, French Patois can be widely heard. The island's turbulent history is reflected in its archaeology, which ranges from Neolithic remains to Royalist castles and concrete defences constructed by the occupying forces during the Second World War.

Sewage treatment on the island could be improved. There are three main outfalls and several minor discharges; there is, however, very little contamination from industrial sources in the island's sewage. Fortunately, a number of the island's beaches do posses water of good quality and we are happy to feature these in the *Guide*.

Jersey was established as an independent state over 700 years ago. When King John lost the island to France in 1204, the islanders chose to remain loyal and to this day they are subject to the Crown, although not governed by Parliament.

Jersey is famous for its coastline. It has an excellent record for water quality and has one of the most comprehensive sewage treatment programmes in Britain, which includes screening, settlement, activated sludge oxidation, secondary treatment and ultraviolet disinfection. Beaches on Jersey are cleaned daily by hand and by machine. Dogs are banned from the beaches in Jersey between 10.30am and 6pm from 1 May to 30 September, with fines for those not complying. Jersey takes real care of its beaches and with 55 kilometres of coastline ranging from high cliffs to sweeping bays and bathing waters which meet the highest EC standards, it must count as a prime destination for the discerning beach lover. An added attraction this year is a flourishing Marine Conservation Society local group.

*The Channel Islands*

# PEMBROKE BAY
## Guernsey
*OS Ref: 340837*

Pembroke Bay, also known as L'Ancresse Bay, is a large, almost unbroken expanse of sandy beach in a horseshoe-shaped gulf. It is sheltered from the winds from most directions, making it ideal for serious sunbathers and for beach games.

### Water Quality
No routine sewage discharge has been identified.

### Bathing Safety
Bathing is safe at any state of the tide. Lifebuoys are provided at key access points and trained first-aiders are in attendance.

### Litter
The beach is cleaned daily by hand during the summer, and four days a week in winter. There are dog litter bins in the car parks. Recycling bins are also provided. Dogs are banned from the beach between May and September.

### Access
At the northernmost tip of the island La Route de L'Ancresse leads to L'Ancresse Common and the Bay. There are slipways at both ends of the bay and a number of steep flights of steps lead down to the beach at points along the sea wall.

### Parking
Adequate parking exists immediately above the beach.

### Public Transport
Buses run regularly to and from St Peter Port. Please check with the bus companies for timetable enquiries: Guernsey Bus Ltd, Tel. 01481 724677; Island Coachways Ltd, Tel. 01481 720210.

### Toilets
At both ends of the beach.

### Food
Food is available from the kiosks at both ends of the beach. Two adjacent hotels serve bar lunches and evening meals.

### Seaside Activities
Swimming and windsurfing are restricted to separate zones of the beach. A windsurfing and Hobie Cat school also offers equipment for hire. There is a fine 18-hole links golf course on L'Ancresse Common, immediately behind the beach.

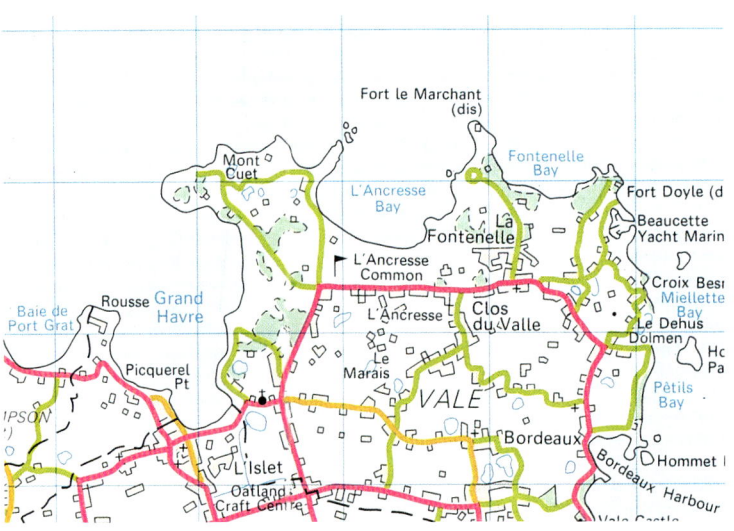

### Wet Weather Alternatives.
Rousse pre-Martello Tower (Fortress Guernsey site), Oatlands craft centre, Guernsey candles, Guernsey freesia centre, Model World, and a Koi fish farm.

### Wildlife and Walks
L'Ancresse Common is the site of several neolithic burial chambers as well as pre-Martello towers dating from Napoleonic times.

*Despite the clouds, Pembroke Bay boasts one of the best sunshine records in Guernsey.*

### Tourist Information
Guernsey Tourist Information Bureau, North Esplanade, St Peter Port, Guernsey, GY1 3AN.
Tel. 01481 723552

**Track Record PGGG**
Not EU designated.

# HAVELET BAY
## Guernsey
*OS Ref: 339778*

South of St Peter Port lies the town beach, Havelet Bay. It has a foreshore of shingle, with sand exposed at low tide, and numerous rock pools. The bay is overlooked by the house of the French poet and novelist, Victor Hugo.

**Water Quality**
No routine sewage discharge has been identified.

**Bathing Safety**
The bay offers safe swimming, with lifebuoys located at key points.

**Litter**
The beach is cleaned by hand five days a week during the summer and three days a week in winter. Litter bins are located along the beach, with a dog litter bin at La Vallette. Dogs are not banned from this beach.

**Access**
Havelet Bay is situated next to the South Esplanade, St Peter Port. The beach is accessible via two slipways at either end.

**Parking**
Car parking along the beach front.

**Public Transport**
The island's main bus terminus is situated about five minutes walk away.

**Toilets**
Toilet facilities are located nearby.

**Food**
Available from neighbouring cafés and restaurants.

**Seaside Activities**
Swimming; waterskiing; diving school nearby.

**Wet Weather Alternatives.**
Aquarium, La Vallette underground military museum, Victor Hugo's house, Castle Cornet & maritime museum, shopping centre, Guernsey museum & art gallery and Beau Sejour leisure centre.

**Wildlife and Walks**
The footpaths of the east coast are accessible from here: several cliff paths lead along shaded woodlands to other bays along the coast. Leaflets are available from the Tourist Information Bureau.

**Tourist Information**
Guernsey Tourist Information Bureau, North Esplanade, St Peter Port, Guernsey, GY1 3AN. Tel. 01481 723552

**Track Record PGPG**
Not EU designated.

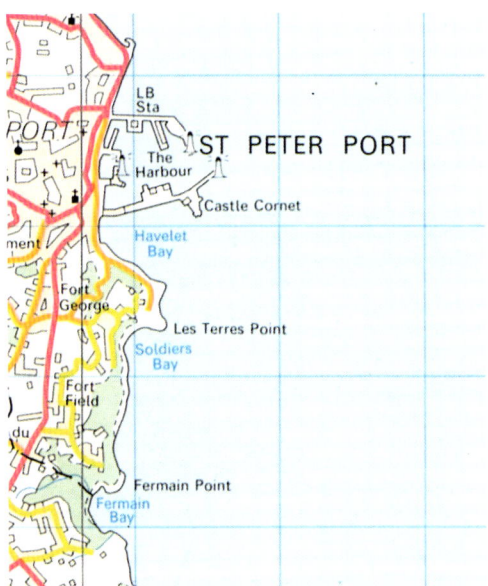

# FERMAIN BAY
## Guernsey
*OS Ref: 336761 (see map opposite)*

The most popular east coast beach, and one of the prettiest on the island, is sheltered from all except east winds. A pebble bank at the top of the beach gives way to firm sand at low tide. The beach is reached via a steep winding road down a beautiful valley (which towards the end is closed to all except essential traffic) or by following the cliff paths.

### Water Quality
No routine sewage discharge has been identified.

### Bathing Safety
The bay offers safe swimming, with lifebuoys located adjacent to slipway.

### Litter
The beach is cleaned by hand daily during the summer, and three to four days a week in the winter. Litter bins are located at strategic points along the beach. Dogs are banned between 1 May and 30 September inclusive. Dog litter bins are provided for the rest of the year.

### Access
Fermain Bay is situated off Fort Road towards La Favorita Hotel and is well signposted. Access is by foot from Le Chalet hotel car park or the cliff path from St Peter Port.

### Parking
Parking is at the top of the access road near Le Chalet hotel.

### Public Transport
Buses run regularly to and from St Peter Port.

### Toilets
Toilet facilities are located nearby.

### Food
Food is served from a café near the slipway; various hotels at the top of the access road serve lunch and other meals.

### Seaside Activities
Good bathing and snorkelling.

### Wet Weather Alternatives.
Saumarez Manor, a stately home with large gardens, model railway and public exhibitions.

### Wildlife and Walks
Fermain is two miles along the cliffs from St Peter Port and forms part of the largest nature conservation site on the island. Many walks criss-cross the south and east coasts and a leaflet is available from the Tourist Information Bureau.

### Tourist Information
Guernsey Tourist Information Bureau, North Esplanade, St Peter Port, Guernsey, GY1 3AN. Tel. 01481 723552

### Track Record PGPG
Not EU designated.

# PETIT BÔT BAY
## Guernsey
*OS Ref: 305749*

This popular bay lies at the foot of two wooded valleys. The beach is pebbled at high tide, but at low tide a large expanse of sand is exposed.

**Water Quality**
No routine sewage discharge has been identified.

**Bathing Safety**
The bay offers safe swimming, with lifebuoy facilities located nearby.

**Litter**
The beach is cleaned by hand daily during the summer, and three to four days a week in winter. Litter bins are located along the beach. Dogs are banned from the beach between 1 May and 30 September inclusive. Dog litter bins are provided for the rest of the year.

**Access**
A clearly signposted road leads down to the beach just before the airport. A

*The Channel Islands*

slipway makes the beach easily accessible.

 **Parking**
At the top of the beach.

**Public Transport**
Buses run regularly to and from St Peter Port. Please check with the bus companies for timetable enquiries: Guernsey Bus Ltd, Tel. 01481 724677; Island Coachways Ltd, Tel. 01481 720210.

 **Toilets**
Toilet facilities are located nearby.

 **Food**
A kiosk serves food and refreshments.

**Seaside Activities**
Good bathing and snorkelling.

**Wet Weather Alternatives.**
German occupation museum; Bruce Russell gold & silversmiths; Guernsey woodcarvers; strawberry farm; bird gardens; St Saviours underground tunnel; Sula gallery; Guernsey clockmakers.

 **Wildlife and Walks**
Icart Point, one of a number of rocky headlands on the south coast, is a short walk to the east and has some of the finest views on the island. Leaflets on cliff walks are available from the Tourist Information Bureau.

**Tourist Information**
Tourist Information Desk, States Airport. Tel. 01481 37267

**Track Record PGPG**
Not EU designated.

*Petit Bôt is typical of the tiny coves and rugged bays of Guernsey's remote south coast.*

# SAINT'S BAY
## Guernsey
*OS Ref: 323747*

A beautiful sandy cove with pebble bank at the top of the beach, surrounded by protective cliffs. A fisherman's landing is situated within the bay.

### Water Quality
No routine sewage discharge has been identified.

### Bathing Safety
The bay offers safe swimming, with lifebuoy facilities located nearby.

### Litter
The beach is cleaned by hand daily during the summer, and three to four days a week in the winter. Litter bins are located throughout the beach. Dogs are allowed on the beach at all times; there are no dog litter bins.

### Access
Well signposted from the centre of St Martin's village. A slipway offers easy access to the beach.

### Parking
Some car parking is available on the steep road to the beach and on the fisherman's landing.

### Public Transport
Buses run regularly to and from St Peter Port. Please check with the bus companies for timetable enquiries: Guernsey Bus Ltd, Tel. 01481 724677; Island Coachways Ltd, Tel. 01481 720210.

### Toilets
Toilet facilities are located nearby.

### Food
A café and hotels in the area serve refreshments and meals.

### Seaside Activities
Good bathing and snorkelling.

### Wet Weather Alternatives.
Moulin Huet pottery, Saumarez Manor & Model Railway, German underground hospital, Little Chapel, Sula Gallery and Guernsey Clockmakers.

### Wildlife and Walks
Leaflets on cliff walks are available from the Tourist Information Bureau.

### Tourist Information
Guernsey Tourist Information Bureau, North Esplanade, St Peter Port, Guernsey, GY1 3AN. Tel. 01481 723552

### Track Record ~~~G
Not EU designated.

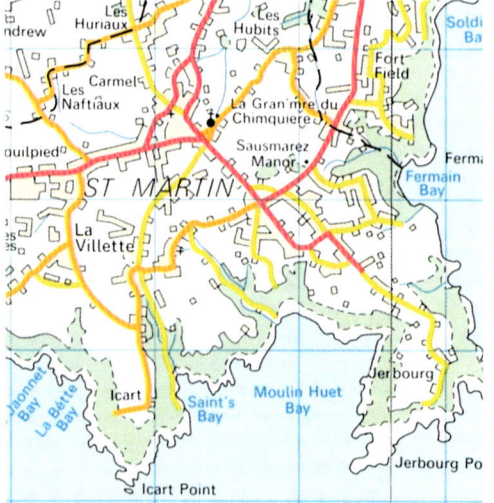

*The lush greenery on the surrounding cliffs gives Saint's Bay a tropical apperance.*

*The Channel Islands*

# PORTELET BAY
## Guernsey
### OS Ref: 245760

This enclosed sandy bay offers safe swimming and this makes it a popular choice for families. The slipway running from the road to the sand also makes it easily accessible for disabled visitors. In addition to the attractions of the beach and the small fishing harbour, the many rockpools offer great potential for those who enjoy exploring. The south coast footpath is also easily reached from this bay.

**Water Quality**
No routine sewage discharge has been identified.

**Bathing Safety**
Lifebuoys are located at key access points. The bay offers safe swimming.

**Litter**
The beach is cleaned four times a week in the winter and daily during the summer. There are a number of litter bins in the surounding area. Dog litter bins are situated at the top of the beach.

**Access**
Follow the west coast road to the south-western tip of the Island, or via La Route de Pleinmont from the airport. The beach is reached via a slipway from the car park.

**Parking**
There is adequate parking.

**Public Transport**
Buses run regularly to and from St Peter Port. Please check with the bus companies for timetable enquiries: Guernsey Bus Ltd, Tel. 01481 724677; Island Coachways Ltd, Tel. 01481 720210.

**Toilets**
Toilet facilities are located at the top of the beach.

**Food**
A kiosk near the car park sells snacks. Meals are served at a nearby hotel.

**Wet Weather Alternatives.**
Fort Grey maritime museum; Coppercraft & Guernsey Pearl Centre; Coach House art gallery; Edda Aschmann silk studio; Le Planel Dolls.

**Wildlife and Walks**
The South Coast Footpath network is easily accessible. The Pleinmont Headland walk is within a few miles of the beach.

**Tourist Information**
Tourist Information Desk, States Airport. Tel. 01481 37267

**Track Record ~~GG**
Not EU designated.

*A Martello tower presides over Portelet Bay's foreshore where the sandy beach is exposed at low tide.*

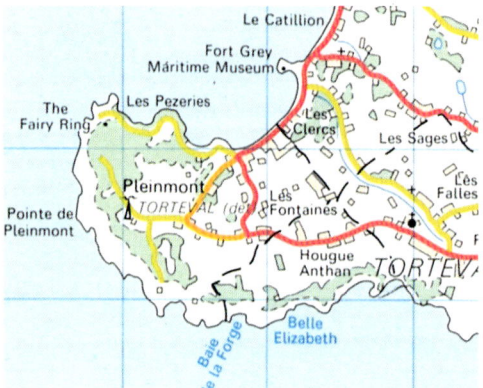

*The Channel Islands*

# VAZON BEACH
## Guernsey
*OS Ref: 285798*

This wide crescent of lovely clean sand gets the sun from early morning until sunset. It is exposed to the wind from most directions, which is bad news for sunbathers but often creates excellent conditions for surfers.

**Water Quality**
No routine sewage discharge has been identified.

**Bathing Safety**
Generally safe, but observe the warning flags. Lifebuoys are provided at key access points.

**Litter**
The beach is cleaned by hand daily during the summer and three to four days a week in the winter. Litter bins are located throughout the beach, and recycling facilities are provided. Dogs are banned between May and September, with dog litter bins provided for the rest of the year.

**Access**
Situated on the west coast between Fort Hommet Headland and Richmond, the beach is reached via flights of steps,

with slipways at either end providing access for disabled visitors.

**Parking**
Plenty of space at two large car parks to the north end of the bay.

**Public Transport**
Buses run regularly to and from St Peter Port. Please check with the bus companies for timetable enquiries: Guernsey Bus Ltd, Tel. 01481 724677; Island Coachways Ltd, Tel. 01481 720210.

**Toilets**
Well maintained toilets are located at both ends of the beach, with facilities for disabled visitors at the north end of the bay, near the terminus car park.

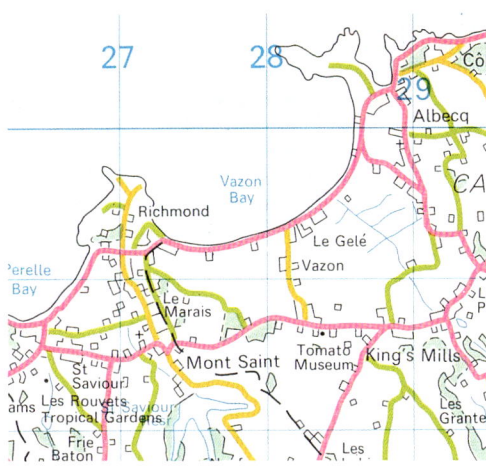

**Food**
There are two beach kiosks and a few restaurants and hotels in the area.

**Seaside Activities**
Good bathing, surfing, windsurfing and canoeing. Activities are clearly zoned.

**Wet Weather Alternatives.**
Fort Hommet German gun casement (Fortress Guernsey Site), Guernsey Tomato Centre, Pamela Dorey art studio, Brooklands Farm Agricultural Implement Museum.

**Wildlife and Walks**
Fort Hommet Headland at the north end of the beach is a Nature Conservation Area, of interest both historically and for its wildlife.

**Tourist Information**
Guernsey Tourist Information Bureau, North Esplanade, St Peter Port, Guernsey, GY1 3AN. Tel. 01481 723552

**Track Record GGGG**
Not EU designated.

*The long and sandy Vazon Beach is especially popular with surfers.*

# COBO BAY
## Guernsey
*OS Ref: 296806*

This long sandy beach is divided into sections by rocky outcrops of pink granite. There is good bathing, snorkelling and windsurfing to be had here.

### Water Quality
No routine sewage discharge has been identified.

### Bathing Safety
Generally safe, with lifebuoy facilities provided at slipways.

### Litter
The beach is cleaned by hand daily during the summer, and four or five days a week in the winter. Litter bins are located throughout the beach, and recycling facilities are provided. Dogs are banned from May to September, with dog litter bins provided for the rest of the year.

### Access
Cobo Bay is situated on the west coast road and is well signposted. Steps from the car park allow access to the beach, with disabled access via the slipways.

### Parking
Good parking facilities are available at the beach.

### Public Transport
Buses run regularly to and from St Peter Port. Please check with the bus companies for timetable enquiries: Guernsey Bus Ltd, Tel. 01481 724677; Island Coachways Ltd, Tel. 01481 720210.

### Toilets
Toilets are located nearby.

### Food
A kiosk and nearby hotels provide snacks, hot meals and refreshments.

### Seaside Activities
Good bathing, snorkelling and windsurfing in designated zones.

### Wet Weather Alternatives.
Activity World; Butterfly Farm; Telephone Museum. The Guernsey Folk Museum, run by the National Trust of Guernsey, is about half a mile to the south.

### Wildlife and Walks
Saumarez Nature Trail and Le Guet Pine Forest; leaflets are available from the Tourist Information Bureau.

### Tourist Information
Guernsey Tourist Information Bureau, North Esplanade, St Peter Port, Guernsey, GY1 3AN.
Tel. 01481 723552

**Track Record PPPG**
Not EU designated.

# PORT SOIF BAY, WEST COAST
## Guernsey
*OS Ref: 305819 (see map opposite)*

Port Soif is an almost entirely circular bay which offers shelter from virtually any wind. It has a foreshore area of wonderful fine, dry sand.

### Water Quality
No routine sewage discharge has been identified.

### Bathing Safety
Generally safe, with lifebuoys at the key access points and a trained first-aider in attendance.

### Litter
The beach is cleaned by hand daily during the summer, and four times a week in the winter. Litter bins are located throughout the beach. Dogs are banned from May to September, with dog litter bins provided for the rest of the year.

### Access
Located on the coast road in the north-west of the island. Access points include steps and a slipway for the disabled.

### Parking
There is plenty of parking immediately above the beach.

### Public Transport
Buses run regularly to and from St Peter Port. Please check with the bus companies for timetable enquiries: Guernsey Bus Ltd, Tel. 01481 724677; Island Coachways Ltd, Tel. 01481 720210.

### Toilets
Well maintained toilets are available.

### Food
A kiosk exists at the top of the beach.

### Seaside Activities
Good bathing.

### Wildlife and Walks
The Port Soif Nature Trail lies just behind the beach, with a leaflet available from the Tourism Information Bureau. Port Soif Common is a Site of Nature Conservation.

### Tourist Information
Guernsey Tourist Information Bureau, North Esplanade, St Peter Port, Guernsey, GY1 3AN. Tel. 01481 723552

### Track Record PGGG
Not EU designated.

*Wildflowers thrive in the dunes behind the beach at Port Soif.*

# ST OUEN'S BAY (LE BRAYE and WATERSPLASH)
## Jersey
*OS Ref: 565514*

St Ouen, the longest beach in Jersey spanning nearly the full length of the island's west coast, is locally refered to as 'the five mile road', for obvious reasons. It is a superb surfing beach and has been the venue for international surfing competitions. The vast sandy beach is backed by Les Mielles conservation area. Situated to the south of the bay is La Rocco Tower which can be visited at low tide but is cut off at mid to high tide.

### Water Quality
No routine sewage discharge has been identified.

### Bathing Safety
The same conditions which make this beach popular with surfers can also make it dangerous for all but the strongest swimmers: observe the safety flags at all times. There is safety cover from the island's main beach guard headquarters between late April and the end of September, to which emergency telephones at seven locations are directly linked.

### Litter
The beaches are cleaned and the litter bins emptied daily between May and September. Dogs are only allowed on the beach if kept on a lead; dog waste must be deposited in the bins provided.

### Access
There are numerous points of access to the beach by means of slipways and steps.

### Parking
There are many car parks along the length of the bay, some of which have been specially designed to blend in with the surrounding dune area by partially enclosed grass banks.

### Public Transport
Bus number 12a runs to St Ouen's Bay.

### Toilets
Numerous toilet facilities along the bay also include facilities for disabled visitors, who can gain access to cubicles across the island by obtaining a Radar key at the Town Hall.

*The Channel Islands*

### Food
Cafés, beach kiosks and licensed restaurants.

### Seaside Activities
Surfing instruction and equipment hire; nine-hole municipal golf course at Les Mielles together with a driving range and crazy golf course.

### Wet Weather Alternatives
Frances le Sueur Centre and Kempt Tower Interpretation Centre.

### Wildlife and Walks
La Mielle de Morville, adjacent to Kempt Tower, is a pleasant place to walk and the starting point for longer circular routes. The sand dunes support over 400 plant species, including wild orchids, sand crocus and autumn squill. The Les Mielles

*So shallow is the slope of the beach at St Ouen's Bay that the water is almost out of sight at low tide.*

Conservation Area is designated to conserve and protect the dune area together with surrounding flora and fauna. There are weekly guided walks around Les Mielles every Thursday afternoon between May and September.

### Tourist Information
Jersey Tourism,
Liberation Square, St Helier, Jersey.
Tel. 01534 500700

**Track Record**
**Le Braye GGGG**
**Watersplash ~~GG**
Neither beach is EU designated.

# BEAU PORT
## Jersey
*OS Ref: 579479 (see map overleaf)*

This beautiful and secluded sandy bay on Jersey's southern coast has a beach sheltered by rocky headlands on either side, and provides safe bathing in a quiet location.

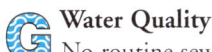

### Water Quality
No routine sewage discharge has been identified.

### Bathing Safety
There is no beach guard cover but bathing is relatively safe.

### Litter
The beach is cleaned and the bins emptied daily between May and September. Dogs are only allowed on the beach if kept on a lead; dog waste must be deposited in the bins provided.

### Parking
There is a car park at the top of the path to the beach.

### Public Transport
The number 12 bus stops about 10 minutes' walk away.

### Toilets
There are no toilet facilities at the beach.

### Food
A refreshment van visits the car park occasionally.

### Seaside Activities
Swimming.

### Wet Weather Alternatives
None at this beach.

### Wildlife and Walks
A path runs along the headland offering spectacular views of the bay. Beauport is home to the Dartford Warbler.

### Tourist Information
Jersey Tourism,
Liberation Square, St Helier, Jersey.
Tel. 01534 500700

### Track Record GGGG
Not EU designated.

*Beau Port provides a quieter and more relaxing alternative to the much larger St Brelade's Bay next door.*

# ST BRELADE'S BAY
## Jersey
*OS Ref: 585486*

The most popular beach on Jersey is particularly favoured by families due to the large range of activities available. A large expanse of white sand is backed by a sea wall and a landscaped promenade which is floodlit in the evenings.

### Water Quality
No routine sewage discharge has been identified.

### Bathing Safety
Professional beach guard cover is provided between May and September. The safe bathing area is designated by red and yellow flags.

### Litter
The beach is cleaned and the bins emptied daily between May and September. Dogs are only allowed on the beach if kept on a lead; dog waste must be deposited in the bins provided.

### Access
There is a ramp designed specifically for beach users with impaired mobility. In addition to this access, there is another slip at the east end of the bay, together with numerous sets of steps.

### Parking
Car parking is situated nearby.

### Public Transport
Bus number 14 stops near the beach.

### Toilets
Toilets at four different sites include facilities for the disabled user.

### Food
There are many restaurants and cafés.

### Seaside Activities
Swimming, windsurfing, canoe hire, boat trips, fun boats for hire.

### Wet Weather Alternatives
Fisherman's Chapel, Red Houses shopping centre and Quennevais sports & leisure centre.

### Wildlife and Walks
The less energetic walker will appreciate the promenade that runs the length of the bay.

### Tourist Information
Jersey Tourism, Liberation Square, St Helier, Jersey.
Tel. 01534 500700

**Track Record PGPG**
Not EU designated.

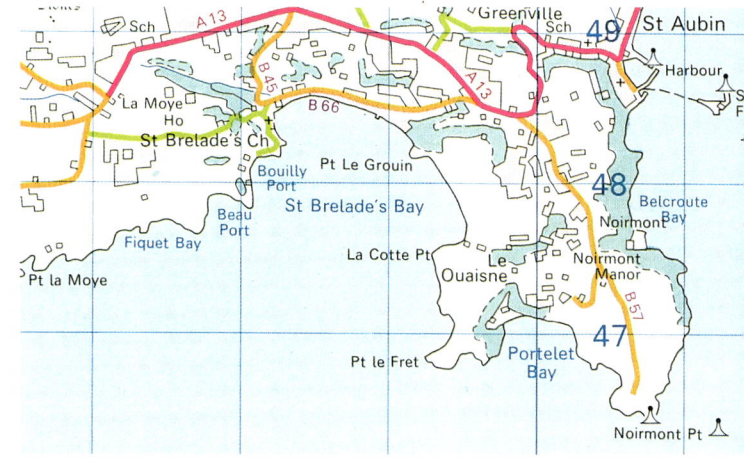

# PORTELET
## Jersey
*OS Ref: 600471 (see map opposite)*

This quiet, sandy beach in a sheltered bay on the south-west coast of the island is popular with the locals. Wooded hills rise steeply from the rocky foreshore and an islet in the bay is fully encircled at half to full tide.

### Water Quality
No routine sewage discharge has been identified.

### Bathing Safety
Generally safe. There is no beach guard cover.

### Litter
The beaches are cleaned and the litter bins emptied daily between May and September. Dogs are only allowed on the beach if kept on a lead; dog waste must be deposited in the bins provided.

### Access
Access is via steep steps with bench seating at three points and not suitable for people with limited mobility.

### Parking
There is parking on the headland.

### Public Transport
Bus number 12.

### Toilets
There are no toilet facilities at the beach.

### Food
A café sells food and beach goods. There is an inn and restaurant adjacent to the car park.

### Seaside Activities
Deck chairs, windbreaks and canoes are for hire. This is a particularly good area for exploring rockpools.

*Just offshore at Portelet lies the Ile au Guerdain, crowned by an 18th-century Martello tower.*

### Wet Weather Alternatives
St Aubin Harbour is a short drive away and has many attractive shops and restaurants.

### Wildlife and Walks
A path leads to Noirmont Point from where there are particularly fine views of the bay and across to St Helier.

### Tourist Information
Jersey Tourism, Liberation Square, St Helier, Jersey. Tel. 01534 500700

**Track Record GGGG**
Not EU designated.

# HAVRE DES PAS
## Jersey
*OS Ref: 658476*

Havre des Pas is a town beach of pebbles and sand. A promenade runs the length of the beach, ideal for walking away the sunny days of summer.

**Water Quality**
No routine sewage discharge has been identified.

**Bathing Safety**
Bathing is relatively safe; there is no lifeguard cover. A bathing pool is patrolled by trained personnel.

**Litter**
The beach is cleaned regularly and litter bins are situated at strategic points. Dogs must be kept on a lead between 10.30am and 6.00pm during the summer, and their owners are responsible for safely disposing of waste.

**Access**
Havre des Pas can be found on the main road leading from St Helier in an easterly direction. A slipway leads down to the beach from the adjacent pavement.

**Parking**
There is no parking at the beach.

**Public Transport**
The beach is within walking distance of St Helier, and is served by the number 18 bus.

**Toilets**
Toilets are located by the bathing pool area, and these include facilities for disabled visitors.

**Food**
A café situated within the bathing pool area sells light refreshments.

**Seaside Activities**
Swimming and fun boat hire.

**Wet Weather Alternatives.**
Close to Fort Regent leisure centre, Howard Davis Park and the main shopping centre of St Helier.

**Wildlife and Walks**
Promenade walk.

**Tourist Information**
Jersey Tourism, Liberation Square, St Helier, Jersey. Tel. 01534 500700

**Track Record ~~PG**
Not EU designated.

*Havre des Pas provides safe bathing within walking distance of Jersey's historic capital, with the added attraction of a supervised pool.*

# GROUVILLE BAY
## Jersey
*OS Ref: 710501*

Grouville, known more familiarly as Gorey, is a stretch of golden sands which sweeps south along the Royal Bay of Grouville to La Roque Point, winter home to large numbers of migrating waders such as turnstones and sanderlings.

### Water Quality
No routine sewage discharge has been identified.

### Bathing Safety
Bathing is relatively safe. There is no lifeguard cover.

### Litter
The beach is cleaned regularly and litter bins are situated at strategic points. Dogs must be kept on a lead between 10.30am and 6.00pm during the summer, and their owners are responsible for safely disposing of waste.

### Access
Grouville can be found by taking the A3 and then the A4 coastal route from St Helier. Access to the beach is adjacent to the Royal Jersey Golf Club.

### Parking
There are parking facilities at the beach.

### Public Transport
Served by bus numbers 1, 1a, and 1b.

### Toilets
Toilets including facilities for disabled visitors are situated nearby.

### Food
There is a refreshment outlet on the beach, a mobile concessions van on the slipway and restaurants in Gorey Harbour.

### Seaside Activities
Swimming, windsurfing and sailing.

### Wet Weather Alternatives.
Mont Orgueil Castle, Jersey Pottery, shops at Gorey Pier.

### Wildlife and Walks
A common adjacent to the golf links runs parallel to the beach providing excellent walks.

### Tourist Information
Jersey Tourism, Liberation Square, St Helier, Jersey. Tel. 01534 500700

**Track Record PPPG**
Not EU designated.

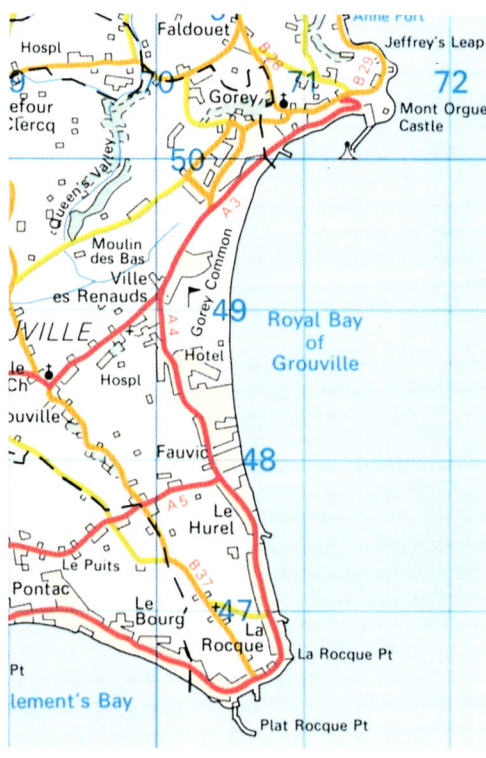

# GREEN ISLAND
## Jersey
*OS Ref: 675462*

Just a ten-minute drive from the town of St Helier, Green Island is a fine sandy beach with outcrops of rock, and a very popular beach with the locals. Low tide reveals dozens of rockpools and also permits access to the island across the sand. Don't get cut off when the tide comes in!

**Water Quality**
No routine sewage discharge has been identified.

**Bathing Safety**
Bathing is considered generally safe. There is no lifeguard cover.

**Litter**
Litter is collected daily during the summer season. Dogs are allowed, but must be kept on a lead between 10.30am and 6.00pm from May to September.

**Access**
One main slipway to the beach.

**Parking**
Adjacent to beach.

**Public Transport**
The no. 1 bus serves the bay.

**Toilets**
Toilet facilities are at the car park.

**Food**
A café serving light refreshments is situated at the beach.

**Seaside Activities**
Swimming.

**Wet Weather Alternatives**
St Helier with its shops and other facilities is a short drive away. Samares Manor is also conveniently located.

**Wildlife and walks**
Green Island is too popular for shore birds to be evident, but a walk east towards Le Hocq Point reveals an excellent habitat for many unusual species.

**Tourist Information**
Jersey Tourism,
Liberation Square, St Helier, Jersey.
Tel. 01534 500700

**Track Record GGGG**
Not EU designated.

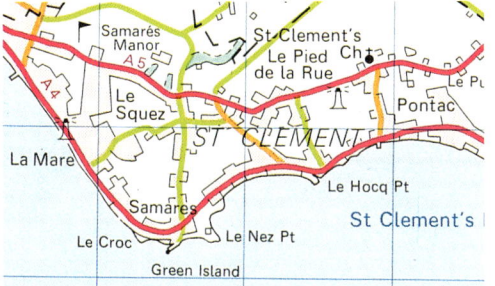

# L'ARCHIRONDEL
## Jersey
*OS Ref: 712517*

This is a quiet shingle beach situated on the east coast of the island, in an area often used by canoeing clubs. The beach is unspoilt and offers spectacular views of St Catherine's Breakwater.

### Water Quality
No routine sewage discharge has been identified.

### Bathing Safety
Bathing is considered safe with care. There is no lifeguard cover.

### Litter
Litter is collected daily during the summer season. There is no dog ban, but dogs must be kept on a lead from 10.30am to 6.00pm from May to September.

### Parking
There is a small car park specifically for the beach and the café.

### Public Transport
Served by bus no. 20.

*Superb bathing in a quiet, sheltered bay is the main attraction of L'Archirondel.*

### Toilets
These are situated in the car park.

### Food
The café next to the beach has both indoor and outdoor seating and serves a wide range of food and drink.

### Seaside Activities
Swimming.

### Wildlife and Walks
There are barbecue and picnic areas close to L'Archirondel offering fine views of the breakwater.

### Tourist Information
Jersey Tourism, Liberation Square, St Helier, Jersey.
Tel. 01534 500700

### Track Record GGGG
Not EU designated.

# BOULEY BAY
## Jersey
*OS Ref: 670547*

Situated on Jersey's north coast, Bouley Bay presents a sweeping shingle beach, set against cliffs to the north and the west. Bathing here is relatively safe, providing care is taken, as there is no lifeguard cover. The site is popular with those who choose to submerge themselves completely in the water, as a school providing training in scuba-diving operates from the bay. The alternative for those not wishing to get wet is one of a number of fine cliff walks which depart from the path above the beach.

### Water Quality
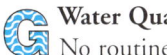
No routine sewage discharge has been identified.

### Bathing Safety
Bathing is recommended for strong swimmers only. There is no lifeguard cover.

### Litter
Litter is collected daily during the summer season. There is no dog ban, but dogs must be kept on a lead from 10.30am to 6.00pm from May to September.

### Access
The bay is reached by a winding road leading down from Trinity.

### Parking
Limited parking is available at the bay.

### Public Transport
The no. 4 bus serves the bay during the summer months only.

### Toilets
There are toilet facilities on site.

### Food
A café serves light refreshments.

### Seaside Activities
Swimming (with caution); a diving school operates from the bay.

### Wildlife and Walks
There is a cliff walk from the bay leading to the pretty fishing port of Rozel.

### Tourist Information
Jersey Tourism, Liberation Square, St Helier, Jersey.
Tel. 01534 500700

**Track Record ~~GG**
Not EU designated.

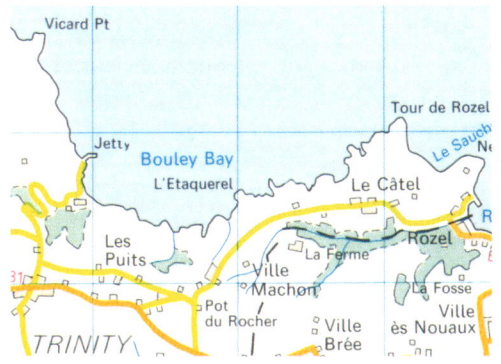

# GREVE DE LECQ
## Jersey
*OS Ref: 583555 (see map opposite)*

This sandy horseshoe-shaped bay has outcrops of flat rocks exposed at low tide. The beach is particularly popular with locals and the jetty at the end of the bay is used by local fishermen.

**Water Quality**
No routine sewage discharge has been identified.

**Bathing Safety**
Bathing is relatively safe, with care. Note that the beach shelves considerably at low tides. There is no lifeguard cover.

**Litter**
The beach is cleaned regularly and litter bins are provided. Dogs must be kept on a lead between 10.30am and 6.00pm during the summer, and owners are responsible for disposing of waste.

**Access**
From St Helier take the A2, then the A12, then the B65 to the beach. There is one main slipway, and two sets of steps.

**Parking**
There are two car parks at the beach.

**Public Transport**
The no. 9 bus serves the beach.

**Toilets**
Toilets, including facilities for the disabled visitor, are situated at the beach.

**Food**
There are two cafés and a licensed restaurant along the promenade.

**Seaside Activities**
Swimming.

**Wet Weather Alternatives.**
Large outdoor children's play area at the popular inn 'Moulin de Lecq.'

**Wildlife and Walks**
Cliff walk to Devil's Hole or Plemont Bay; follow the signs from the main car park.

**Tourist Information**
Jersey Tourism, Liberation Square, St Helier, Jersey. Tel. 01534 500700

**Track Record PGPG**
Not EU designated.

# PLEMONT
## Jersey
*OS Ref: 561566*

Plémont is a must for anyone who enjoys actively exploring the coastline. Situated on the north-west coast of the island, it is famed for its fascinating caves which have been eroded into the cliffs surrounding the bay. A cliff path also leads west from the bay to Les Landes. If relaxing on the beach is more to your taste, you will find a pleasing expanse of sand at Plémont. It should be noted that the beach is reached via a series of steep wooden steps and this may make access difficult for less mobile visitors.

### Water Quality
No routine sewage discharge has been identified.

### Bathing Safety
Bathe with caution. There is lifeguard cover from May until September. The beach is covered at high tide.

### Litter
The beach is cleaned regularly and litter bins are provided. Dogs must be kept on a lead between 10.30am and 6.00pm during the summer, and owners are responsible for removing waste.

### Access
The beach is reached by way of a narrow and winding road. The sands are accessed from the road by means of steep steps.

### Parking
There is limited parking at the top of the steps leading to the beach.

### Public Transport
The no. 8 bus serves the beach during the summer months.

### Toilets
Toilets are situated adjacent to the café.

### Food
A café at the top of the steps sells refreshments.

### Seaside Activities
Swimming. The bay features a number of caves which may be explored.

### Wet Weather Alternatives.
Plémont Candlecraft is situated immediately off the main road on the approach to the beach.

### Wildlife and Walks
A cliff path leads west to Les Landes and connects with a network of other walks serving the island. There is a resident colony of puffins and razorbills at Plémont.

### Tourist Information
Jersey Tourism,
Liberation Square, St Helier, Jersey.
Tel. 01534 500700

### Track Record GPGG
Not EU designated.

## Channel Islands

| RATING | NAME | TRACK RECORD | SEWAGE OUTLET | REMARKS |
|---|---|---|---|---|
|  | **GUERNSEY** |  |  |  |
| G | **Pembroke Bay** 340837 | PGGG | 🟩 | FEATURED |
| G | **Havelet Bay** 339778 | PGPG | 🟩 | FEATURED |
| G | **Fermain** 336761 | PGPG | 🟩 | FEATURED |
| ~ | **Moulin Huet Bay** 330751 | ~~~~ | 🟩 | 🟨 Fascinating rock formations. |
| G | **Petit Bot Bay** 305749 | PGPG | 🟩 | FEATURED |
| G | **Saint's Bay** 323747 | ~~~G | 🟩 | FEATURED |
| G | **Portelet Bay** 245760 | ~~GG | 🟩 | FEATURED |
| P | **L'Eree** 255780 | PGGP | 🟩 | 🟨 🗑 |
| ~ | **Perelle Bay** 264788 | ~~~~ | 🟩 | 🟨 ⛰ Poor for swimming. |
| G | **Vazon Beach** 285798 | GGGG | 🟩 | FEATURED |
| G | **Cobo Beach** 296806 | PPPG | 🟩 | FEATURED |
| ~ | **Saline Bay** (Grandes Rocques) 299816 | ~~~~ | 🟩 | 🟨 🟧 |
| G | **Port Soif Bay** (West Coast) 305819 | PGGG | 🟩 | FEATURED |
| ~ | **Grand Havre** 326826 | ~~~~ | 🟩 | 🟨 Edged by granite outcrops. |
| P | **Ladies Bay** | ~~~P | 🟩 | 🟨 |
|  | **JERSEY** |  |  |  |
| G | **St Ouen's Bay** Le Braye 565514 | GGGG | 🟩 | FEATURED |
| G | Watersplash 565514 | ~~GG | 🟩 | FEATURED |
| G | **Beauport** 579479 | GGGG | 🟩 | FEATURED |
| G | **St Brelade's Bay** 585486 | PGPG | 🟩 | FEATURED |
| G | **Portelet** 600471 | GGGG | 🟩 | FEATURED |
| G | **Havre des Pas** 658476 | ~~PG | 🟩 | FEATURED |
| G | **Grouville** 710501 | PPPG | 🟩 | FEATURED |
| G | **Green Island** 675462 | GGGG | 🟩 | FEATURED |
| G | **L'Archirondel** 712517 | GGGG | 🟩 | FEATURED |
| G | **Bouley Bay** 670547 | ~~GG | 🟩 | FEATURED |

🟨 Sand  Shingle 🟦 Pebbles ⛰ Rocks 🟫 Mud ❓ No information supplied

*Channel Islands*

| RATING | NAME | TRACK RECORD | SEWAGE OUTLET | REMARKS |
|---|---|---|---|---|
| G | **Greve de Lecq**<br>583555 | PGPG | 🟩 | FEATURED |
| G | **Plémont**<br>561566 | GPGG | 🟩 | FEATURED |

🟩 No discharge identified   ⬆ Improvements planned   ⁇ Insufficient information to feature   🗑 Cleaned regularly

# Britain's Coastline Under Threat

The problem of sewage pollution is the reason this book was first written and sewage is still a widespread problem in Britain. Of the hundreds of millions of litres of sewage pumped into the sea around our coasts every day, much is either raw or receives little treatment. Even where full secondary treatment is applied, raw sewage can still enter the sea via combined sewer overflows (CSOs), particularly after heavy rainfall. A mixture of domestic waste water, cleaning agents, industrial and trade effluent, solid litter and storm water, it will typically contain human wastes, sanitary protection, condoms, bathroom wastes, engine oils, fat balls from domestic and trade kitchens, and a range of heavy metal contaminants (mercury, lead, cadmium, arsenic, copper) from trade effluent, detergents and road surface run-off. It also contains viruses and bacteria that can cause disease in humans.

To catch a potentially fatal illness from sewage whilst bathing is rare, but studies show that the chances of contracting other less serious illnesses are considerable. It is entirely possible to contract an ear, nose and throat infection from the water, or to suffer diarrhoea and vomiting after swimming. Each year the Marine Conservation Society receives reports from people who have become ill after bathing, windsurfing or diving in sewage-contaminated waters.

When you consider what sewage is, the harm it can do and how much of it we produce, comprehensive treatment would seem an essential precaution to protect our health and the marine environment. But very little sewage in this country receives adequate treatment, despite the fact that the technology exists to make sewage pollution a thing of the past.

## Sewage Treatment – the Ideal and the Reality

Comprehensive sewage treatment consists of a number of stages. The first and most basic, PRELIMINARY TREATMENT, attempts to remove the larger solids – the plastics, nappies and all manner of debris that finds its way into the sewers – by a coarse filtering process called SCREENING. The material screened out is largely un-recyclable and is disposed of in landfill sites or possibly incinerated. Preliminary treatment may also include a process called MACERATION, which is roughly equivalent to putting the sewage through an enormous food blender. Preliminary treatment is a starting point for the proper treatment of sewage, but in no sense adequate in itself.

The sewage now contains 200-500 milligrams per litre of suspended solids, and the next stage, known as PRIMARY TREATMENT, allows these to settle out. Several hours of settlement before the effluent is passed on removes 50-60% of suspended solids and up to 50% of bacteria and viruses – though the effluent will still contain many millions of these organisms per litre – and produces a large volume of sludge.

The SLUDGE generated by sewage treatment should not be regarded as a waste to be disposed of as quickly and as cheaply as possible, but as a resource to be exploited. There are many options for its useful utilisation: sludge can be used for the production of fertilisers, soil conditioners, peat substitutes, methane production for electricity generation and even oil production. An innovative scheme piloted in Britain by Wessex Water using the Swiss Combi process has shown that a commercial fertiliser can be made from sewage sludge.

Ideally, following primary treatment, the effluent should then be subjected to SECONDARY TREATMENT to stimulate biological activity and reduce the oxygen demand of the sewage; there are several ways of achieving this, all of which generate more sludge which has to be dealt with. Secondary treatment removes 90-95% of suspended solids, 80-90% of

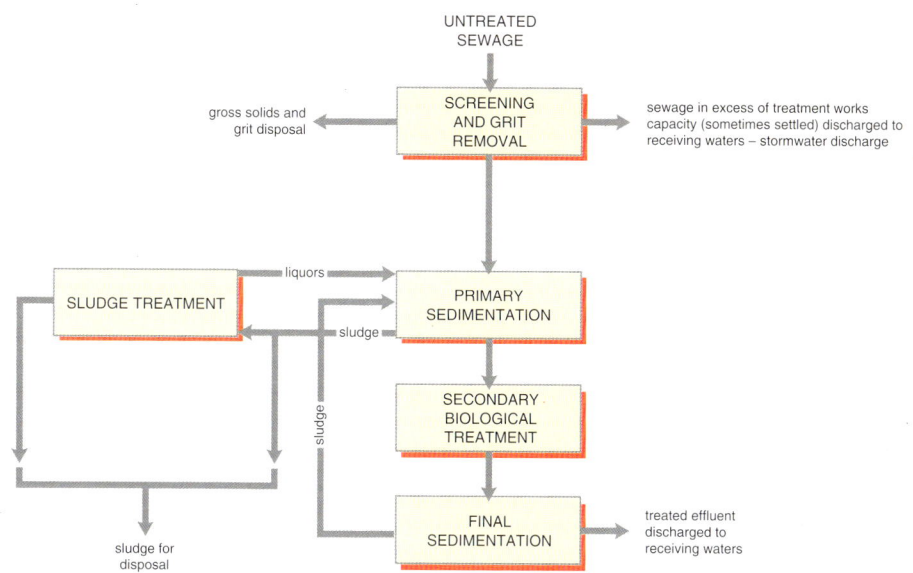

*Figure showing full sewage treatment.*

the oxygen demand, 75-99% of bacteria and about 50% of heavy metal contamination.

In some cases, though it is still rare in the UK, TERTIARY TREATMENT may be used to reduce the nitrogen and phosphorus levels in the effluent in order to lessen its fertiliser effect, in a process known as nutrient stripping. This is important since nutrients in sewage (For instance phosphorus and nitrogen) can severely disrupt aquatic ecosystems, causing algal blooms and the death of marine and freshwater life. Inputs of nutrients from agricultural practices also contribute greatly to the problem of nutrient enrichment and pollution.

Irrespective of what type of treatment it receives, sewage is discharged to sea via OUTFALL PIPES which vary widely in how far they extend into the sea. Many discharge a few metres below low water mark and some may even discharge above the level of low tide. In the past, LONG SEA OUTFALLS were seen as the solution to contaminated bathing waters, but discharging the sewage as far out to sea as possible is no longer regarded as a substitute for treating it. There are significant environmental problems with raw sewage wherever it is discharged. Sewage slicks from long sea outfalls may be washed back towards beaches by the wind, waves and currents and result in unpleasant encounters for swimmers and sailors.

At whatever stage the effluent is discharged to sea, whether raw or treated, CHEMICAL DISINFECTION may sometimes be applied. The chemicals used may include sodium hypochlorite, peracetic acid or ozone. None of these has been adequately tested to ensure its safety with regard to marine life and human health when used in the disinfection of sewage. There is also doubt about the efficacy of these methods in destroying the pathogens in the effluent: the use of chemical disinfection may be lulling us into a false sense of security by removing the indicators of sewage pollution, while the disease-causing agents remain. It is widely agreed that chemical disinfection is no substitute for comprehensive sewage treatment.

PHYSICAL DISINFECTION, using ultraviolet light systems or possibly ultra-filtration methods, appears to present no threat to the marine environment, since these treatments are non-additive. The use of such systems is increasing in popularity among a number of organisations within the water industry, including Dwr Cymru/Welsh Water and Jersey's water authority. The latter recently won an environmental tourism award for its use of ultraviolet disinfection in ensuring the quality of the island's bathing water.

In addition to the routinely discharging outfalls, OVERFLOW or STORM WATER OUTFALLS (CSOs) come into use during storms and times of heavy rainfall – on average around 10 times a year – as treatment works become overloaded and sewers fill up. These discharge effectively untreated sewage, sometimes directly on to the beach above low water mark.

**Sewage Pollution and the Law**

There is no doubt that improvements are being made to sewage treatment systems, but progress is slow. The water service companies have control over the vast majority of discharges (with the exception of a small number of privately operated sewage outfalls) and as such are directly responsible for this mass pollution of our seas. Although some companies are investing millions of pounds in long term improvement projects, others appear to be doing the bare minimum required of them by law. There are three main pieces of legislation that refer directly to the discharge of sewage to the sea:

Under the Water Resources Act (1991), the Environment Agency must give consents to discharge to the water companies for each sewage discharge to sea. These consents should be designed to protect the waters into which the sewage is being discharged, but in practice the system does not always work effectively and can be undermined by political considerations. Some coastal discharges regularly breach their consents and many simply do not have numerical consents, meaning that no limit is put on the volume of raw sewage that goes down the outfall pipe to the sea. In Scotland, these aspects are regulated by the Scottish Environment Protection Agency (SEPA). Unfortunately, it is not just sewage that ends up in the sewers: surprisingly, legislation still allows industrial discharges – containing heavy metals, industrial detergents, dyes and oils – to be made to sewers, often in very inappropriate circumstances where sewage works are not designed to cope with these pollutants.

A European Union law, the Urban Waste Water Treatment Directive (91/27/EEC), seeks to make secondary treatment the standard minimum level of treatment throughout the Union for all coastal sewage discharges serving populations of more than 10,000 people and estuarine discharges which serve more than 2,000 people. Full implementation of the Directive will undoubtedly produce an improvement in the coastal waters around Britain, although the Directive itself is not entirely satisfactory, since numerous smaller outfalls are not covered and primary treatment will be considered adequate for large outfalls if the coastal waters are deemed to be areas of high natural dispersion, or declared 'less sensitive'. The Government has already declared many areas of the British coast less sensitive, including some of Britain's most polluted estuaries which should be treated as priorities for a long overdue clean-up. The UK appears to be as enthusiastic for the Urban Waste Water Treatment Directive as it was for the Bathing Water Directive when it was first introduced: at that time the Government identified only 27 bathing waters in Britain, while land-locked Luxembourg declared 34! The Urban Waste Water Treatment Directive must be fully implemented. The Marine Conservation Society will be looking closely at the scientific basis on which areas are classified as being of 'high natural dispersion' and will take any issues arising to the European Commission and Parliament, as well as ensuring that the UK meets its promises to comply with the deadlines in the Directive.

The Government has recently ratified Annex IV of the International Agreement on the Prevention of Pollution from Ships (MARPOL). This makes it an offence to discharge sewage without treatment from ships, large boats and yachts. Holding tanks on ships are to be encouraged so that sewage can be discharged to land-based sewage treatment works at ports and harbours, and reception facilities here will have to be upgraded to cope: there is no point in bringing raw sewage back to land if the land-based solution is simply to pump it back to sea via an outfall pipe. While it appears to be a step in the right direction, many countries with large shipping fleets do not recognise the MARPOL agreement and continue to discharge raw sewage at sea. Boat owners wanting guidance on how to deal with sewage should contact either the Royal Yachting Association who publish the *Clean Code for Boat Owners and Users*, or the British Marine Industries Federation who have produced *Navigate with Nature*. Both addresses are in the Appendix

## Bag It and Bin It

Despite the ready availability of the technology for adequate treatment, and in the face of the legal framework designed to protect the seas, the simple fact is that much of the sewage from coastal populations discharges without any form of treatment. As long as this is the case, those who use the sea will have to put up with sewage-related debris on the beach and will continue to fall ill from sewage-related disease. We can all help reduce the problem of debris; it is a simple matter to dispose of such things as sanitary items, condoms or cotton buds in the bin rather than down the lavatory pan. Simply bagging and binning waste products and using less plastic will help clean up beaches and bathing waters in the short term, as well as making the water companies' treatment of sewage easier. Less debris in the sewers would mean fewer blockages in the pipes and fewer breakdowns in the treatment machinery.

## The Problem of Marine Litter

The dropping of litter in public places – which includes beaches – was outlawed by the Environmental Protection Act (1990). Tourist or recreational litter has been identified as a major contributor to the litter on Britain's beaches, consistently making up over 20% of litter recorded in Beachwatch surveys.

Littering at sea is also illegal under Annex V of MARPOL which prohibits the dumping of plastics overboard from ships and boats. This is regulated in British waters under the Merchant Shipping (Prevention of Pollution by Garbage) Regulations (1988), but these appear to be less than adequately enforced. The Marine Conservation Society is campaigning hard for a tightening of the law governing the disposal of waste from ships at sea, with the aim of making ships deposit garbage when in port as well as making ports responsible for providing adequate receiving facilities. The Reader's Digest Beachwatch beach cleaning campaign, run by the Marine Conservation Society, has consistently found that over 50% of all debris collected is plastic. And as compliance with Annex V is voluntary and has been ratified by only 65% of the gross shipping tonnage of the world fleet, it is all too clear that many ships are still dumping rubbish over the side at sea.

We can all play a part in helping to reduce the amount of litter on our beaches by ensuring that we put our own rubbish in the bin or take it with us.

## Other Forms of Marine Pollution

Catastrophic incidents of accidental pollution – the horrific oil spills in the Gulf, the Exxon Valdez incident in Alaska in 1989 and, closer to home, the MV Braer in Shetland in 1993 and Sea Empress in 1996 – seem to occur with such regularity that one could be forgiven for thinking that they were the main source of pollution. In fact accidents of this sort account for only around 10% of the oil that finds its way into the sea each year. The remainder comes from routine losses and controllable discharges. Continual chronic oil pollution is caused by the deliberate and illegal flushing of tanks at sea by the bulk oil carriers, by spills at on-shore and off-shore installations during the loading and unloading of tankers and by pipeline fractures which allow tonnes of oil to escape into the seas each year: evidence of this can be seen in the form of sticky tar found on many beaches in the UK. Oil has a clogging and smothering effect on marine life. About 60% of dead seabirds found around the UK are oiled; in the English Channel this figure rises to a shocking 75%. Marine life under the water is also at risk; under certain weather conditions the heavier fractions of an oil slick sink to the sea bed, forming an impenetrable barrier under which bottom-dwelling animals and plants die.

**How you can help**
A visit to the beach should be a pleasant experience but so often it is not. If you encounter dirty water, sewage, litter or other dangerous items, then COMPLAIN ABOUT IT. If the authorities are unaware of the problem, or think that people don't care then the situation will not improve. There are various bodies – local and national government agencies, the water companies and others – to whom you should report pollution. The responsibility of these relevant bodies are outlined below

If you have a complaint about a beach you visit, please write to us at the Marine Conservation Society and tell us about it. Tell us all you can. We can advise you on what action to take with the relevant authorities, although we cannot act on individual cases. The more we know about the problems around our coast, the more we can do about them.

While local authorities in England and Wales do not have responsibility for sewage treatment, they are charged with keeping beaches free of litter and safe from dangerous items – chemical drums and canisters, for instance – that may be washed up. If you find items that you think may be dangerous, call the Environmental Health Department straight away. If you discover items such as chemical drums or explosives washed up on the shore, the Coastguard should be contacted immediately. On no account touch whatever it is you find. You should also register a complaint if the beach is littered with drink cans, sewage related debris, plastic bottles or discarded fishing nets, as the local authority has a legal duty under the Environmental Protection Act 1990 to keep public places clean and free of litter. Local Authorities are also responsible for any pollutant considered to be a statutory nuisance – including sewage-related debris – so tell the Environmental Health Department where the litter is and ask that they clear it up.

These departments are also responsible for publishing the results of bathing water monitoring at the EU designated beaches in line with government policy. The results should be displayed on posters in a clear, easy-to-understand form at the beaches themselves. If they are not, contact the Chief Environmental Health Officer to ask why. Ask that posters be displayed near the beach showing up-to-date water quality monitoring results.

Although tourism departments have no direct responsibility for pollution on the beach or out at sea, they stand to lose if a resort gets a bad name as a result of beaches polluted with sewage or litter. They are usually quick to act, passing on any complaints they receive to the relevant agency and ensuring that the problem is addressed.

The Environment Agency is responsible for water quality and pollution incidents in England and Wales. The Agency also carries out the routine monitoring of designated bathing waters, whose results are used to assess compliance with the EC Bathing Waters Directive. Non-designated bathing waters may also be sampled either by the Agency, or in some cases by local authorities or the water companies. The Scottish Environmental Protection Agency

(SEPA) in Scotland and the Department of the Environment for Northern Ireland (DoE-NI) carry out broadly similar roles to those of the Environment Agency in England and Wales, as does the States Board of Administration in the Channel Islands.

If you come across an instance of sewage pollution in the water, or if there is evidence of other pollution, contact the local Environment Agency or SEPA office or call the Agency's Emergency Hotline on 0800 807060. They should be able to investigate the pollution and may be able to track down and prosecute the polluter. The address and telephone numbers of the responsible bodies are given in the Appendix. If you do report an incident, keep up the pressure until you are satisfied the Agency has taken appropriate remedial or legal action.

The private water companies are responsible for the operation of coastal sewage works and outfall pipes around the coastline of England and Wales. In Scotland, Northern Ireland and the Channel Islands, the water industry has not been privatised. The private water service companies in England and Wales are currently investing large sums of money in a range of improvement projects, but pressure must be maintained to ensure they do their job properly.

Some areas can boast tremendous success stories, but in others the scheduled improvements are not progressing at an adequate rate. The implementation of the EU Urban Waste Water Directive will see much investment and a corresponding improvement in sewage treatment. Some water companies, however, openly state that they will do only the bare minimum required of them by the Directive and the Environment Agency. Neither their interests nor ours are served by this sort of short-sighted attitude. Not only is sewage pollution deeply unpleasant and potentially hazardous to people using the beach, it is also extremely costly, in the loss of revenue from tourism, in the drain on precious health service resources treating bathing-related illnesses, in legal settlements and fines arising from the increasing number of legal actions filed against the water companies and in the direct cost of picking up sewage-derived rubbish from our beaches.

If you come across a pipe discharging raw sewage near a beach, write to your local water company. Ask whether they consider it acceptable to discharge raw sewage to sea. Ask them for information about the beaches you visit. How many outfall pipes are there and how many people do they serve? To what level is sewage treated prior to discharge? Are the outfalls clearly marked giving details of how much sewage is discharged or what sort of treatment is received? If there is no treatment, why not? Ask for details of improvement schemes. What level of treatment will be provided by the improvement scheme and when will it be completed? Make sure that your local MP and MEP know your views. It is important that they know there is great interest and concern about coastal pollution – that there are 'votes in sewage.'

If we ask no questions, we will get no answers. The authorities responsible for coastal water quality must be held to their promises of improvements. We must let them know how we feel. They must understand that we value our precious coastal environment and want money spent on safeguarding it. The capacity of the sea to assimilate waste is not infinite and we cannot continue to use it as a dustbin. International agreements and national legislation, often only enacted following the application of campaigning pressure by groups like the Marine Conservation Society and private individuals such as yourself, must be enforced and rigorously policed. We still have a long way to go if we are to enter the next century with clean seas. Giving support to organisations such as the Marine Conservation Society is one way to voice your concern. It is the duty of us all to speak out, to report pollution where we see it and to complain about the degradation of a unique resource. The longer we remain silent, the worse it will become. We must act now to safeguard our coast for ourselves and for other creatures and to ensure that it remains a fitting heritage for future generations.

# Useful Address and Contacts

**THE MARINE CONSERVATION SOCIETY**
9 Gloucester Road,
Ross-on-Wye,
Herefordshire, HR9 5BU.
Tel: 01989 566017
Fax: 01989 567815

**THE COASTGUARD**
The Coastguard is available to help anyone in danger at sea or on the beach. If you think someone needs help,
**DON'T HESITATE – DIAL 999**
and ask for the Coastguard.

**THE ENVIRONMENT AGENCIES**
The England & Wales Environment Agency Water Pollution Hotline :
0800 807060
(Free*fone* number)

**Head Office**
Rio House,
Waterside Drive,
Aztec West,
Almondsbury,
Bristol, BS12 4UD.
Tel: 01454 624400
Fax: 01454 624409

**London Office**
Eastbury House,
30-34 Albert Embankment,
London, SE1 7TL.
Tel: 0171 820 0101
Fax: 0171 820 1603

**North East Region**
Rivers House,
21 Park Square South,
Leeds, LS1 2QG.
Tel: 01532 440191
Fax: 01532 461889

**Anglian Region**
Kingfisher House,
Goldhay House,
Orton Goldhay,
Peterborough, PE2 0ZR.
Tel: 01733 371811
Fax: 01733 231840

**South West Region**
Manley House,
Kestrel Way,
Exeter, EX2 7LQ.
Tel: 01392 444000
Fax: 01392 444238

**Southern Region**
Guildbourne House,
Chatsworth Road,
Worthing, BN11 1LD.
Tel: 01903 820692
Fax: 01903 821832

**North West Region**
12 Richard Fairclough House,
Knutsford Road,
Warrington, WA4 1HG.
Tel: 01925 653999
Fax: 01925 6415961

**Thames Region**
Kings Meadow House,
Kings Meadow Road,
Reading, RG1 9DQ.
Tel: 01734 535000
Fax: 01734 500388

**Welsh Region**
Rivers House
St Mellons Business Park
Cardiff, CF3 OLT.
Tel: 01222 770088
Fax: 01222 798555

**Scottish Environmental Protection Agency**
Erskine Court
The Castle Business Park
Stirling, FK9 4TR.
Tel: 01786 457700
Fax: 01786 446885

**Environment & Heritage Service, Northern Ireland**
Environment Protection Division, Calvert House,
23 Castle Place,
Belfast, BF1 1FY.
Tel: 01232 254754
Fax: 01232 254700

**WATER COMPANIES**
**Anglian Water**
Anglian House,
Amsbury Road,
Huntingdon, PE18 6NZ.
Tel: 01480 443000

**Northumbrian Water**
Abbey Road,
Pity Me,
Durham, DH1 5FJ.
Tel: 0191 383 2222

**North West Water**
Dawson House,
Great Sankey,
Warrington, WA5 3LW.
Tel: 01925 234000

**Severn Trent Water**
2297 Coventry Road,
Sheldon,
Birmigham, B26 3PU.
Tel: 0121 722 4000

**Southern Water**
Southern House,
Yeoman Road,
Worthing, BN13 3NX.
Tel: 01903 264444

**South West Water**
Peninsula House,
Rydon Lane,
Exeter, EX2 7HR.
Tel: 01392 446688

**Thames Water**
Nugent House,
Vastern Road,
Reading, RG1 8DB.
Tel: 01734 591159

**Wessex Water**
Wessex House,
Passage Street,
Bristol, BS2 OJQ.
Tel: 01179 929 0611

**Yorkshire Water**
West Riding House,
67 Albion Street,
Leeds, LS1 5AA.
Tel: 0113 244 8201

**Welsh Water/Dwr Cymru**
Plas-y-Ffynnon,
Cambrian Way,
Brecon,
Powys, LD3 7HP.
Tel: 01874 623181

**North of Scotland Water Authority**
Caledonian House,
63 Academy Street,
Inverness, IV1 1LU.
Tel: 01463 245400

**East Of Scotland Water Authority**
Pentland Gait,
567 Calder Road,
Edinburgh, EH11 4HJ.
Tel: 0131 445 4141

**West of Scotland Water Authority**
419 Balmoral Road,
Glasgow, G22 6NU.
Tel: 0141 355 5180

## LEISURE AND RECREATION CONTACTS

**The National Trust**
36 Queen Anne's Gate,
London, SW1H 9AS.

**National Trust For Scotland**
5 Charlotte Square,
Edinburgh, EH2 4DU.

**The Manx National Trust**
The Manx Museum,
Douglas,
Isle of Man.

**The Isles Of Scilly Environmental Trust**
Hamewith, The Parade,
St Mary's,
Isles Of Scilly, TR21 0LP.

**Jersey – Beach Quality**
Public Services,
States Offices,
South Hill,
Jersey.

**Guernsey – Beach Quality**
States Board Of Administration,
Sir Charles Frossard House,
La Charroterie,
St Peter Port,
Guernsey, GY1 1FH.

**The Royal Yachting Association**
RYA House,
Romsey Road,
Eastleigh,
Hampshire, SO5 4YA.

**British Marine Industries Federation**
Meadlake Place,
Thorpe Lea Road,
Egham,
Surrey, TW20 8HE.

**British Sub Aqua Club**
Telford's Quay,
Ellesmere Port,
South Wirral, L65 4FY.

**PADI International Limited**
Unit 6, Unicorn Park,
Whitby Road,
Bristol, BS4 4EX.

**Royal Life Saving Society**
Moutbatten House,
Studley,
Warwickshire, B80 7NN.

# INDEX

**Aberaeron**
  North Harbour 194
  North Beach 194
  North of Groynes 194
  South Beach 194
Aberafan 198
  East 198
  Margam Sands 198
Aberarth 194
Abercastle 195
Aberdaron Beach 192
Aberdeen, Footdee 150
Aberdour
  Harbour 149
  Silversands 149
Aberdyfi 193
Abereiddy Bay *166*, 195
Aberffraw Bay 191
Abergele, Towyn 190
Pensarn 190
Abermawr *164*, 195
Aberporth
  Slip 195
  Traeth-y-Dyffryn 194
Aberystwyth
  Harbour 193
  North *160*, 193
  South *160*, 193
  Tanybwich Beach 194
Abrahams Bay 79
Achiltibuie 152
Achmelvich 152
Achnahaird 152
Afan (Port Talbot) 198
Afon Wen 192
Ainsdale 131
Aldeburgh 112
Aldingham 131
Allhallows 111
Allonby
  South 130
  West Winds 130
Alnmouth 127
Amble 127
Amlwch 191
Anderby Beach 124
Annan 154
Anstey's Cove *46*, 80
Anstruther 149
Applecross 152
Arbroath
  Victoria Park 150
Archirondel *238*, 242
Ardrossan 153
Ardwell Bay 154
Arnside 131

Askam-in-Furness 130
Ayr 153

**Babbacombe**
  Beach *46*, 80
Baglan (Neath) 198
Ballycastle 210
Ballyherbert 210
Ballyholme 210
Ballywater 210
Balmedie *142*, 150
Bamburgh 127
Banff Bridge 150
Bantham 79
Barafundle Bay 196
Bardsea 131
Barmouth 193
Barmston 124
Barrow-in-Furness 130
Barry
  Little Island Bay 199
  Watch House Bay 199
Beacon Cove *43*, 80
Beadnell Bay 127
Beaumaris 191
Beau Port *230*, 242
Bedruthan Steps 74
Beer 81
Beesands *38*, 79
Belhaven Beach 148
Bembridge 82
Bendricks Beach 199
Benllech 191
Benone
  Magilligan Strand *202*, 210
Berrow
  North 72
  South 72
Berwick-upon-Tweed 127
Bexhill (Egerton Park) 109
Bigbury-on-Sea
  North 78
  South 78
Birling Gap 109
Bispham 131
Black Rock Sands 193
Blackhall 126
Blackpool Sands *40*, 79
Blackpool
  North Pier 131
  Central (Lost Children's Post) 131
  South Pier 131
Blackwaterfoot 153
Blue Anchor 72

Blundell Sands 132
Blyth 127
Bognor Regis Pier 108
Bognor Regis 108
Borth Wen:
  Rhoscolyn 191
Borth 193
Boscastle 74
Botany Bay 111
Bouley Bay *239*, 242
Bournemouth:
  Alum Chine 82
  Bournemouth Pier 82
  Boscombe Pier 82
  Durley Chine *66*, 82
  Fisherman's Walk *66*, 82
  Southbourne 82
Bovisand Bay 78
Bow (or Vault Beach),
  Gorran Haven 77
Bowleaze 81
Bracelet Bay 198
Bracklesham Bay 108
Brandy Cove 197
Branscombe 81
Braystones 130
Brean 72
Bridlington:
  South 124
  North 124
Brighouse Bay 154
Brightlingsea *103*, 112
Brighton:
  Kemp Town 109
  Palace Pier 109
Broad Haven:
  North *172*, 196
  South *180*, 196
Broadsands Beach 79
Broadstairs:
  Broadstairs Beach 111
  East Cliff 111
  Joss Bay *95*, 111
Brodick Bay 153
Broomhill Sands 110
Broughton Bay 197
Broughty Ferry 149
Browns Bay 210
Bude:
  Crooklets *20*, 73
  Sandy Mouth *20*, 73
  Summerleaze 73
Budleigh Salterton 80
Burghead 151
Burnham-on-Sea
  Yacht Club 72

Jetty 72
Burntisland 149
Burry Port Beach East 197
Burton Bradstock 81
Butlins (Heads of Ayr) 153

**Caerfai Bay** 195
Calgary Bay 152
Calshot 108
Camber Sands 110
Cambois
  North 127
  South 127
Camusdarrach 152
Canvey Island 111
Carbis Bay
  Porth Kidney *28*, 75
  Station Beach 75
Carne Beach 77
Carnlough 210
Carnoustie 150
Carradale 152
Carreg Wen 193
Carrick Shore 154
Castlerock 210
Castletown 132
Caswell Bay 197
Cawsand Bay 78
Cayton Bay 125
Cemaes Bay 191
Challaborough *34*, 78
Chapel St Leonards 124
Charlestown 77
Charmouth
  East 81
  West 81
Chesil Cove 81
Christchurch Bay
  (Barton-on-Sea) 108
Christchurch
  Avon Beach 82
  Friars Cliff 82
  Mudeford Sandbank *70*, 82
  Mudeford Quay 82
Church Bay 191
Church Ope Cove 81
Churston Cove 79
Cil Borth 194
Clachtoll 152
Clacton
  Connaught Gardens 112
  Groyne 112
  Coastguard Station 112
Clarach Bay
  North of River 193

252

| | | | |
|---|---|---|---|
| South of River 193 | (Sugary Cove) 79 | **Fairbourne** 193 | Power Station 113 |
| Clashnessie Bay 152 | Dawlish | Fairlie 153 | South 113 |
| Cleethorpes 124 | Coryton Cove 80 | Fall Bay 197 | Greatstone Beach 110 |
| Clevedon | Dawlish Warren 80 | Falmouth | Green Island 237, 242 |
| Bay 72 | Town 80 | Feock Loe Beach 76 | Greenaway Beach 74 |
| Swimming Pool 72 | Daymer Bay 23, 74 | Gyllyngvase 76 | Grève de Lecq 240, 242 |
| Cleveleys 131 | Deal Castle 110 | Swanpool Beach 76 | Gronant 190 |
| Clovelly 73 | Deganwy 190 | Featherbed Rocks 26 | Groomsport 210 |
| Cobo Beach 226, 242 | Denemouth South 126 | Felixstowe | Grouville Bay 236, 242 |
| Cocklawburn Beach 127 | Derbyhaven 132 | North 112 | Gruinard Bay 152 |
| Cold Knap Beach 199 | Dhoon 154 | South 112 | Gullane 148 |
| Coldbackie 151 | Dinas Dinlle 192 | Felpham (Yacht Club) 108 | Gunwalloe Cove 76 |
| Coldingham Bay 148 | Doniford 72 | Fenell Beach 132 | Gurnard Bay 82 |
| Collieston 150 | Doonfoot 153 | Fermain 217, 242 | Gwbert-on-Sea 195 |
| Colwell Bay 82 | Dornoch 151 | Ferryside 197 | |
| Colwyn Bay | Douglas | Ffrith 190 | **Hallsands** 79 |
| Marine Road 190 | Broadway 132 | Filey 125 | Hampton Pier Beach 111 |
| End of Cayley | Palace 132 | Findochty 151 | Happisburgh 113 |
| Promenade 190 | Summerhill 132 | Fisherrow 148 | Harlech 193 |
| Opposite Rhos | Dover Harbour 110 | Flamborough | Harlyn Bay 74 |
| House Hotel 190 | Dovercourt 112 | North Landing 124 | Harrington 130 |
| Combe Martin 73 | Downderry 78 | South Landing 124 | Hartland Quay 18, 73 |
| Compton Bay 82 | Drummore 154 | Fleetwood (Pier) 131 | Hartlepool 125 |
| Constantine Bay 24, 74 | Druridge Bay 127 | Flimby 130 | Harwich (Sailing Club) 112 |
| Conwy Morfa 190 | Dunbar East 148 | Folkestone 110 | Hastings |
| Cooden Beach 109 | Duncansby Head 151 | Font-y-Gary Bay | Bulverhythe 109 |
| Coral Beaches 152 | Dunglass 148 | (Rhoose) 199 | Fairlight Glen 110 |
| Coverack 76 | Dunnet Bay 151 | Formby 132 | Queens Hotel 110 |
| Cowes | Dunoon 152 | Fowey | St Leonard's Beach 110 |
| East 82 | Dunster | Readymoney Cove 78 | Havelet Bay 216, 242 |
| West 82 | North West 72 | Fraisthorpe 124 | Haverigg 130 |
| Crackington Haven 74 | South East 72 | Fraserburgh 150 | Havre des Pas 234, 242 |
| Craig Dwllan 191 | Dunwich 112 | Freshwater | Hayle, The Towans 28, 75 |
| Cramond 148 | Duporth Beach 77 | East 196 | Hayling Island |
| Cranfield Bay 211 | Durdle Door | West 179, 196 | East 90, 108 |
| Crantock 75 | East 56, 82 | Frinton-on-Sea 112 | West 90, 108 |
| Craster 127 | West 56, 82 | | Heacham |
| Crawfordsburn 210 | Dymchurch | **Gailes** 153 | North Beach 113 |
| Cresswell 127 | Dymchurch Beach 110 | Gairloch 152 | South Beach 113 |
| Criccieth Beach 192 | Hythe Road 110 | Gammon Hole 79 | South Beach |
| East 192 | Redoubt 110 | Gansey Bay 132 | (Near River) 113 |
| Crimdon Park 126 | | Gilfach yr Halen 194 | Helen's Bay 210 |
| South 126 | **Carls Dyke** 124 | Girvan 153 | Helensburgh 152 |
| Crinnis Beach | Earlsferry 149 | Glenarm 210 | Hemmick Beach 77 |
| Golf Links 77 | Easington 126 | Goodrington Sands 79 | Hemsby 113 |
| Leisure Centre 77 | East Looe 76 | Goodwick Beach 195 | Hengistbury Head 68, 82 |
| Par Sands 77 | East Quantoxhead 72 | Goodwick Harbour | Herne Bay 111 |
| Cromarty 151 | East Runton 113 | (South) 195 | Hest Bank 131 |
| Cromer 104, 113 | Eastbourne | Goring-by-the-Sea 109 | Heysham 131 |
| Croyde Bay 73 | East of Pier 109 | Gorleston Beach 112 | Highcliffe Castle 82 |
| Cruden Bay 150 | Wish Tower 109 | Gosford Sands 148 | Highcliffe 82 |
| Cuckmere Haven 109 | Eastney 108 | Gourock (West Bay) 152 | Hightown 132 |
| Cullen 150 | Elbury Cove 80 | Grand Havre 242 | Hill Head 89, 108 |
| Cushendun 210 | Elie/Earlsferry Beach | Grange-over-Sands | Holcombe 80 |
| Cwm yr Eglws 195 | 138, 149 | and Kents Bay 131 | Holland-on-Sea 112 |
| Cwmtydu 194 | Embleton Bay 127 | Great Mattiscombe 79 | Hollicombe 80 |
| | Erraid 152 | Great Yarmouth | Holy Island |
| **Dale** 196 | Exmouth 48, 80 | Caister Point 113 | (Lindisfarne) 127 |
| Dalton Burn 126 | Eyemouth 148 | North 113 | Holywell Bay 75 |
| Dartmouth Castle | Eypemouth 53, 81 | Pier 113 | Hope Cove 79 |

253

| | | | | | | | |
|---|---|---|---|---|---|---|---|
| Hopeman | 151 | Leonard's Cove | 79 | Lusty Glaze | 74 | Morfa Aberech | 192 |
| Horden | 126 | Lepe | *88*, 108 | Lydstep Haven | *183*, 196 | Morfa Bychan | 193 |
| Hornsea | 124 | Leven | | Lyme Regis | | Morfa Bychan | |
| Hove | 109 | East | 149 | Church Beach | 81 | (Slipway) | 194 |
| Hunstanton | | West | 149 | Cobb | *50*, 81 | Morfa Nefyn | 192 |
| Boat Ramp | 113 | Leysdown-on-Sea | *98*, 111 | Monmouth Beach | 81 | Mortehoe | |
| Hunstanton Beach | | Limeslade Bay | 198 | Lynmouth | 73 | Barricane Bay | 73 |
| | *106*, 113 | Limpert Bay | 199 | Lytham St Anne's | 131 | Rockham Bay | 73 |
| North Beach Sailing | | Little Haven | 196 | | | Mossyard | 154 |
| Club | 113 | Little Hogus | 76 | **Mablethorpe** | 124 | Mothecombe | 78 |
| South Beach | 113 | Little Perhaver | 77 | Macrihanish | 152 | Mother Ivey's Bay | 74 |
| Hythe | 110 | Little Quay | | Maen Porth | 76 | Moulin Huet Bay | 242 |
| | | (Cei Beach) | 194 | Maidencombe | *46*, 80 | Mousehole | 76 |
| **Ilfracombe** | | Littlehampton | 108 | Maidens | 153 | Mouthwell Sands | 79 |
| Capstone Beach | 73 | Littlestone-on-Sea | 110 | Manorbier | 196 | Muchalls | 150 |
| Hele Beach | 73 | Lizard, Church Cove | 76 | Marazion and Mounts Bay | | Mundesley | 113 |
| Tunnels | 73 | Llanaber (Dyffryn) | 193 | Heliport | 76 | Murkle Bay | 151 |
| Ingoldmells | 124 | Llanbedrog | 192 | Wherrytown | 76 | Murlough | 210 |
| Instow | 73 | Llandanwg | 193 | Margate | | Musselwick Sands | 196 |
| Inverberie | 150 | Llanddona (Red | | The Bay | 111 | Mwnt | *162*, 195 |
| Inverboyndie | 150 | Wharf Bay) | *156*, 191 | Fulsam Rock | 111 | | |
| Irvine | 153 | Llanddulas | 190 | Marloes Sands | *174*, 196 | **Nairn** | |
| | | Llanddwyn Beach, | | Marsden Bay | 126 | Central | 151 |
| **Jacksons Bay** | 199 | Niwbwrch | *158*, 191, | Marske-by-the-Sea | 125 | East | 151 |
| Jangye Ryne | 76 | Llandudno | | Martin's Haven | *176*, 196 | Nash Point | 199 |
| Jaywick | 112 | North Shore | 190 | Maryport | 130 | Ness Cove | *46*, 80 |
| Jersey Marine | | West Shore | 190 | Mawgan Porth | 74 | Nethertown | 130 |
| Central | 198 | Llanelli Beach | | Meadfoot Beach | *44*, 80 | New Brighton | |
| East | 198 | (Fourth Groyne) | 197 | Menai Straits | | (Harrison Drive) | 132 |
| West | 198 | Llanfairfechan | 190 | Plas Menai | 192 | New Quay | |
| Jurby | 132 | Llangranog | 194 | Porth Dinorwic Sailing | | Harbour | *161*, 194 |
| | | Llanina | 194 | Club | 192 | Traeth Gwyn | 194 |
| **Kames Bay** | 152 | Llanon (Slipway) | 194 | Meols | 132 | Traeth y Dolau | 194 |
| Kennack Sands | 76 | Llanrhystud | 194 | Mevagissey | 77 | Newbiggin | |
| Kessingland | 112 | Llansantffraid | 194 | Mewslade | 197 | (Northumberland) | |
| Kilchattan Bay | 152 | Llanstephan and | | Middleton-on-Sea | 108 | North | 127 |
| Kimmeridge Bay | *58*, 82 | Tywn Estuary | 197 | Milford Beach | 196 | South | 127 |
| Kinghorn | 149 | Llantwit Major Beach | 199 | Milford-on-Sea | *86*, 108 | Newbiggin (Cumbria) | 130 |
| Kingsands Bay | 78 | Llwyngwril | 193 | Mill Bay | 79 | Newcastle | *206*, 210 |
| Kingsdown Beach | 110 | Lodmoor | 81 | Millendreath | 78 | Newgale Sands | *170*, 195 |
| Kinmel Bay | | Lodmoor West | 81 | Millisle | 210 | Newhaven West Quay | 109 |
| (Sandy Cove) | 190 | Longhoughton Steel | 127 | Millom | 130 | Newhaven | 109 |
| Kirk Michael | 132 | Longniddry | 148 | Millport | 153 | Newport Car Park Slip | 195 |
| Kirkcaldy Linktown | 149 | Looe | | Milsey Bay | 148 | Newport Sands | |
| Knot End-on-Sea | 131 | Hannafore | 78 | Minehead | | North | 195 |
| Kynance Cove | 76 | Plaidy | 78 | Terminus | 72 | South | 195 |
| | | Lossiemouth | | The Strand | 72 | Newquay | |
| **L'Eree** | 242 | East | 151 | Minnis Bay | 111 | Fistral Beach | 75 |
| Ladies Bay | 242 | Silversands | 151 | Minster Leas | 111 | Great Western | 74 |
| Ladram Bay | 80 | West | 151 | Moelfre | | Tolcarne Beach | 74 |
| Lamlash Bay | 153 | Low Newton (Newton | | (Treath Lligwy) | 191 | Town Beach | 74 |
| Lamorna Cove | 75 | Haven) | *122*, 127 | Moggs Eye | 124 | Newry Beach, | |
| Lancing (South) | 109 | Lower Largo | 149 | Monifieth | 149 | Holyhead | 191 |
| Langland Bay | 198 | Lowestoft | | Monreith | 154 | Newton Bay | |
| Lansallos Bay | 78 | South Beach | 112 | Montrose | 150 | (Newton Point) | 198 |
| Lantic Bay | 78 | North Beach | 112 | Morar | 152 | Neyland Slip | 196 |
| Largs | 153 | Lulworth Cove | 82 | Morecambe | | Nicholson's Strand | |
| Laxey | 132 | Lunan Bay | 150 | North | 131 | (Kilkeel) | *208*, 211 |
| Lee-on-the-Solent | 108 | Lunderston Bay | 152 | South | 131 | Nigg Bay | 151 |
| Leigh-on-Sea | 111 | Lundin Links | 149 | Moreton | 132 | Nolton Haven, | 196 |

| | | | | |
|---|---|---|---|---|
| Norman's Bay | 109 | Pontllyfni | 192 | (Outer Harbour) 154 | St Dogmaels Slipway 195 |
| North Berwick Bay | 148 | Poole | | Portrush | St George's Pier |
| North Cliffs (Deadman's | | Branksome Chine 82 | Curran Strand 210 | Menai Bridge 191 |
| Cove, Cambourne) 75 | | Lake 82 | Mill Strand 210 | St Helen's 71, 82 |
| Norton | 82 | Rockley Sands 82 | Portstewart 210 | St Ishmael (Kidwelly) 197 |
| | | Sandbanks 65, 82 | Portwrinkle, Freathy 78 | St Ives |
| **Oddicombe** | *46*, 80 | Shore Road 64, 82 | Powfoot 154 | Porthmeor *28*, 75 |
| Ogmore-by-Sea | 199 | Poppit Sands | Praa Sands | Porthminster *28*, 75 |
| Old Portsmouth | | East 195 | East *31*, 76 | Porthwidden *28*, 75 |
| Beach | 108 | West 195 | West *31*, 76 | St Just, |
| Overcome | 81 | Porlock Weir 72 | Prestatyn Centre, 190 | Priest's Cove *30*, 75 |
| Overstrand | 113 | Porreath 75 | Prestwick 153 | St Margaret's Bay 110 |
| Oxwich Bay | 197 | Porscatho 77 | Putsborough Beach *17*, 73 | St Mary's Bay (Devon) 79 |
| | | Port Erin 132 | Pwllheli 192 | St Mary's Bay (Kent) 110 |
| **Padstow** | 74 | Port Eynon 197 | | St Mary's Well Bay 199 |
| Pagham | 108 | Port Isaac 74 | **Ramsey** 132 | St Mawes 77 |
| Paignton | | Port Soderick 132 | Ramsgate | St Mildred's Bay 111 |
| Paignton Sands | 80 | Port Soif Bay *227*, 242 | Ramsgate Beach 111 | St Ouen's Bay |
| Preston Sands | 80 | Port St Mary 132 | Ramsgate Sands *94*, 110 | Le Braye *228*, 242 |
| Palm Bay | 111 | Portelet (Jersey)*233*, 242 | Ravenglass 130 | Watersplash *228*, 242 |
| Parton | 130 | Portelet (Guernsey) | Redcar | Saint's Bay *220*, 242 |
| Patch | 195 | *222*, 242 | Coatham Sands 125 | Salcombe |
| Pathhead Sands | 149 | Porth Beach 74 | Granville 125 | Millbay 37 |
| Pease Sands | *134*, 148 | Porth Colman 192 | Lifeboat Station 125 | North Sands *36*, 79 |
| Peel | 132 | Porth Dafarch 191 | Stray 125 | South Sands 79 |
| Peffersands | 148 | Porth Dinllaen 192 | Reighton Sands *118*, 125 | Saline Bay |
| Pembrey Beach *184*, 197 | | Porth Eileen 191 | Rhos-y-Llan 192 | (Grandes Rocques)242 |
| Pembroke Bay | *214*, 242 | Porth Iago 192 | Rhosneiger 191 | Saltburn-by-the-Sea 125 |
| Penarth | 199 | Porth Nefyn 192 | Rhossili Bay, *186*, 197 | Saltcoats 153 |
| Penbryn | 194 | Porth Nobla 191 | Rhyl 190 | Saltdean 109 |
| Pendine Sands | 197 | Porth Tywyn Mawr 191 | Ringstead Bay 81 | Sandend Bay 150 |
| Pendower Beach | 77 | Porthallow 76 | Rinsety Head 76 | Sandgate |
| Penmaenmawr | 190 | Porthmeor 74 | Roan Head 130 | Sandgate Beach 110 |
| Penryn Bay | 190 | Porthcawl | Robin Hood's Bay 125 | Town Centre 110 |
| Pentewan | 77 | Rest Bay *188*, 198 | Rockcliffe 154 | Sandhead 154 |
| Penzance | 76 | Sandy Bay 198 | Rock Beach 74 | Sandown |
| Perelle Bay | 242 | Trecco Bay *188*, 198 | Roker | Esplanade 82 |
| Perran Sands | 76 | Porthcothan 74 | Blockhouse 126 | Yaverland 82 |
| Perranporth | | Porthcurnick Beach 77 | Whitburn 126 | Sandsend 125 |
| Penhale Sands | *26*, 75 | Porthcurno 75 | Roome Bay | Sandside Bay 151 |
| Village End Beach*26*, 75 | | Porthglais 195 | (Crail) *140*, 149 | Sandwich Bay 110 |
| Petershead Lido *141*, 150 | | Porthleven West 76 | Rosehearty *146*, 150 | Sandwood Bay 151 |
| Petit Bôt Bay | *218*, 242 | Porthluney Cove 77 | Rosemarkie 151 | Sandy Bay (Devon) 80 |
| Pettycur | 149 | Portholland Beach 77 | Row Cove Beach 78 | Sandy Haven *177*, 196 |
| Pevensey Beach | 109 | Porthor 192 | Runswick Bay 125 | Sandyhills 154 |
| Pilling Sands | 131 | Porthoustock 76 | Ryde | Sango Bay/ |
| Pittenweem | 149 | Porthtowan Sandy 75 | East 82 | Balnakeil Bay 151 |
| Plémont | *241*, 242 | Porthpean *32*, 77 | Seaview 82 | Sanna Bay 152 |
| Plymouth Hoe | | Porthwarra 75 | West 82 | Saundersfoot |
| East | 78 | Portkil/Meiklcross 152 | | Amroth 197 |
| West | 78 | Portland Harbour | **St Andrews** | Beach 197 |
| Point of Ayr Lighthouse190 | | Castle Cove 81 | East 149 | Coppet Hall 197 |
| Poldhu Cove | 76 | Sandsfoot 81 | West Sands 149 | Wiseman's Bridge 197 |
| Polkerris | 77 | Portloe 77 | St Anne's, North 131 | Saunton Sands 73 |
| Polpeor | 76 | Portlogan 154 | St Anthony's Head 77 | Scarborough |
| Polperro | 78 | Portmahomock 151 | St Bees 130 | North Beach 125 |
| Polridmouth Beach | 77 | Portmellon 77 | St Brelade's Bay *232*, 242 | South Beach 125 |
| Polstreath | 77 | Portobello, (Sussex) 109, | St Brides Haven 196 | Scatby Beach 113 |
| Polurrian Cove | 76 | Portobello (Lothian) 148 | St Combs *144*, 150 | Scourie 152 |
| Polzeath | 74 | Portpatrick | St Cyrus 150 | Sea Palling 113 |

255

| | | | | | | |
|---|---|---|---|---|---|---|
| Seacliff | 148 | Southgate | | Totland | 82 | West Runton | 113 |
| Seaford | 109 | (Pwlldu Bay) | 197 | Towan Beach | 77 | Westbrook Bay | 111 |
| Seagrove Bay | 82 | Southport | 131 | Towans, The | | Westcliff-on-Sea | 111 |
| Seaham | | Southsea | | Godrevy | 75 | Westgate Bay | 96 |
|   Beach | 126 | (South Parade Pier) | 108 | Hayle, St Ives | 26, 75 | Westhaven | 150 |
|   Remand Home | 126 | Southwick | 109 | Traeth Llyfn | 165, 195 | Weston-super-Mare | |
| Seahorses | 127 | Southwold | 112 | Traeth Llydan | |   Grand Pier | 72 |
| Seamill | 153 | Spittal Quay | 127 | (Broad Beach) | 191 |   Kewstoke Sand Bay | 72 |
| Seascale | 130 | Spittal | 127 | Traeth Penllech | 192 |   Main Beach | 72 |
| Seaton | 81 | Staithes | 125 | Traigh, Arasaig | 152 |   Near Marine Lake | 72 |
| Seaton Beach | 78 | Stevenston | 153 | Trearddur Bay | 191 |   Sanatorium | 72 |
| Seaton Carew | | Stokes Bay, Gosport | 108 | Trebarwith Strand | 74 |   Uphill Slipway | 72 |
|   Centre | 125 | Stonehaven | | Trefor | 192 | Westward Ho! | 73 |
|   North | 125 |   Cowie | 150 | Tregardock | 74 | Weymouth | |
|   North Gare | 125 |   Garron | 150 | Tresaith | 194 |   Central | 54, 81 |
| Seaton Sluice | 127 | Stranraer | | Tresilian Bay | 199 |   Lodmore | 54 |
| Seatown | 52, 81 |   Cockle Shore | 154 | Trevaunance Cove | 75 |   South | 81 |
| Selsey Bill | 93, 108 |   Marine Lake | 153 | Trevone Bay | 74 | Whitburn | 126 |
| Seton Sands | 148 | Strathlene, Buckie | 151 | Treyarnon Bay | 24, 74 | Whitby | 125 |
| Shakespeare Cliff | 110 | Strete Gate Beach | 79 | Troon | | White Park Bay | 210 |
| Shaldon | 80 | Studland | 60, 82 |   North | 153 | White Sands Bay | 136, 148 |
| Shanklin | 82 | Sutherland | |   South | 153 | White Strand | 132 |
| Sheerness | |   Hendon South | 126 | Tunstall | 124 | Whitecliff Bay | 82 |
| (Beach Street) | 100, 111 |   Ryhope South | 126 | Turnberry | 153 | Whitehaven | 130 |
| Shell Bay (Poole) | 62, 82 |   Sunderland Beach | 126 | Tynemouth | | Whitesand Bay | 75 |
| Shell Bay (Fife) | 82 | Sutton-on-Sea | 116, 124 |   Cullercoats | 126 | Whitesands Bay | |
| Sheringham, | 105, 113 | Swanage | |   King Edwards Bay | 126 |   St Davids | 168, 195 |
| Shipload Bay | 73 |   Central | 82 |   Long Sands South | | Whiting Bay | 153 |
| Shoalstone Beach | 41 |   North | 82 | | 119, 126 | Whitley Bay | 126 |
| Shoeburyness | 112 |   South | 82 |   Long Sands North | 126 | Whitmore Bay | 199 |
|   East | 102, 112 | Swansea Bay | | Tyrella Beach | 204, 210 | Whitstable | 111 |
| Shoreham | |   County Hall | 198 | Tywyn | 193 | Widemouth Bay | 22, 73 |
| (Kingston Beach) | 109 |   Knap Rock | 198 | | | Willsthorpe | 124 |
| Siddick | 130 |   Black Pill Rock | 198 | **Upper Largo** | 149 | Winchelsea | 110 |
| Sidmouth | |   The Mumbles | 198 | | | Withernsea | 124 |
|   Jacob's Ladder | 80 |   Mumbles Head Pier | 198 | **Vazon Beach** | 224, 242 | Wittering | |
| Silecroft | 130 |   Sketty Lane | 198 | Ventnor | 82 |   East | 92, 108 |
| Silloth | 130 |   Slip | 198 | | |   West | 92, 108 |
| Silver Bay, | | | | **Walney Island** | | Wonwell Sands | 79 |
|   Rhoscolyn | 191 | **Tal-y-Bont** | 193 |   Biggar Bank | 131 | Woody Bay | 73 |
| Silverknowles | 148 | Talland Bay | 78 |   Sandy Gap | 131 | Woolacombe | |
| Sinclairs Bay | 151 | Tankerton Beach | 111 |   West Shore | 130 |   Village Beach | 16, 73 |
| Skegness | 124 | Tayport | 149 | Walpole Bay | 111 | Worbarrow | 82 |
| Skinburness | 130 | Teignmouth | 80 | Walton-on-the-Naze | 112 | Workington | 130 |
| Skinningrove | 125 | Tenby | | Warkworth | 120, 127 | Worthing | |
| Skipsea Sands | 124 |   North | 196 | Warren, The | 110 |   East | 109 |
| Skrinkle Haven | 182, 196 |   South | 196 | Watchet | 72 |   West | 109 |
| Slapton Sands | 39, 79 | Tentsmuir Sands | 149 | Watcombe Beach | 46, 80 | Wringcliff | 73 |
| Snettisham Beach | 113 | Thorntonloch | 135, 148 | Watergate Bay | 74 | | |
| Soar Mill Cove | 79 | Thornwick Bay | 124 | Welcombe Mouth | 73 | **Yarmouth** | 82 |
| Solent Breeze | 108 | Thorpe Bay | 112 | Wells-next-the-Sea | 113 | Yellowcraig | 148 |
| South Milton Sands | 35, 79 | Three Cliffs Bay | 197 | Wembury | 78 | Ynyslas | |
| South Shields | 126 | Thurlestone North | 35, 79 | Wemyss Bay | 152 |   East Tywyni | |
| Southend-on-Sea | 112 | Thurso | 151 | West Angle Bay | 178, 196 |   Ynyslas Estuary | 193 |
| Southerndown | | Tintagel | 74 | West Bay (West) | 81 |   North | 193 |
| (Dunraven Bay) | 199 | Torcross | 79 | West Kirby | 132 | | |
| Southerness | 154 | Torre Abbey Sands | 42, 80 | West Mersea | 112 | | |